Programming Microsoft® Composite UI Application Block and Smart Client Software Factory

David S. Platt

图书在版编目(CIP)数据

Microsoft Composite UI Application Block 和 Smart Client Software Factory 编程:英文/(美)普拉特著. –上海:上海世界图书出版公司,2009.1

ISBN 978 – 7 –5062 –9171 –2

Ⅰ. M… Ⅱ.普… Ⅲ.主页制作 –程序设计 –英文 Ⅳ.TP393.092

中国版本图书馆 CIP 数据核字(2008)第 180647 号

Microsoft Composite UI Application Block
和 Smart Client Software Factory 编程

[美]大卫·普拉特 著

上海世界图书出版公司出版发行

上海市尚文路 185 号 B 楼

邮政编码 200010

(公司电话:021 –63783016 转发行部)

上海竞成印务有限公司印刷

如发现印装质量问题,请与印刷厂联系

(质检科电话:021 –56422678)

各地新华书店经销

开本:850 ×1168　1/16　印张:14　字数:450 000

2009 年 1 月第 1 版　2009 年 1 月第 1 次印刷

ISBN 978 – 7 –5062 –9171 –2/T · 186

图字:09 –2008 –627 号

定价:168.00 元

http://www.wpcsh.com.cn

http://www.mspress.com.cn

To My Family

Contents at a Galance

Table of Contents

What do you think of this book? We want to hear from you!

Microsoft is interested in hearing your feedback so we can continually improve our books and learning resources for you. To participate in a brief online survey, please visit:

www.microsoft.com/learning/booksurvey

Introduction

The first release of the Composite UI Application Block took place in December of 2005, and the first release of the Smart Client Software Factory in July of 2006. This book might therefore seem to be appearing a bit late in the lifecycle of the software, at least compared to others that I've written. However, CAB and SCSF originally represented not so much a finished product as a living design pattern, and as such underwent rapid iteration and change as a result of the feedback of the earliest developers. The evolution of the *WorkItem* class from a use case to a scoping container, discussed in Chapter 3, is an example of this developer-driven change.

In this book, I've tried to bring you the most current thinking on the state of CAB and SCSF and their usage today. You will find it substantially different than the original articles and documentation. I wrote most of this book just prior to the May 2007 release of the Smart Client Software Factory. After its release, I had to scramble to bring the book up to date by our reserved press time. I didn't manage to get the disconnected operation blocks in, but I did cover its use of WPF. Because the May SCSF contains a smoother tool set, I've written all the code samples based on it. CAB strikes me as much closer to the beginning of its life than it is to the end, so it makes sense to go with the latest and best.

What about Acropolis, which has just released its first CTP version as we go to press? Does it make sense to learn and use CAB when Acropolis is on the way? I think it does. First, Acropolis is an evolutionary based on the principles of CAB, so studying and becoming fluent in the latter will help you transition to the former when the time comes. Second, Acropolis is a large and ambitious undertaking, for which a firm schedule has not yet been announced. I think that there's at least a good year and a half, possibly two years, before it gets into wide circulation. So again, studying and developing with CAB is a good thing to be doing today.

I wrote this book in a different style than my previous books for Microsoft Press. Instead of writing a high-level overview as I've done for those titles, (*Understanding COM+* 1999, *Introducing Microsoft .NET* 2001, 2002, and 2003, and the *Microsoft Platform Ahead*, 2004), this book is a detailed, code-level book. It's organized in a workbook format, each two-page spread discussing a particular topic and usually a short code sample pertaining to the discussion. I've tried to break it down into bite-sized chunks to make it easier for you to swallow. I used this approach successfully in my books on COM (Prentice-Hall, 1996, 1997, and 1999), as a more accessible introduction than users could get from Brockschmidt's comprehensive tome.

Who This Book Is For

I wrote this book for both software architects and programmers. I discuss the high-level architectural concepts alongside the code that you need to write to make those concepts happen. Managers will also benefit from reading it. At the very least, their managees won't be able to pull the wool over their eyes quite so easily.

System Requirements

You'll need the following hardware and software to build and run the code samples for this book:

- Microsoft Windows XP with Service Pack 2, Microsoft Windows Server 2003 with Service Pack 1, or Microsoft Windows 2000 with Service Pack 4

- Microsoft Visual Studio 2005 Standard Edition or Microsoft Visual Studio 2005 Professional Edition

- Microsoft SQL Server 2005 Express (included with Visual Studio 2005) or Microsoft SQL Server 2005

- 600 MHz Pentium or compatible processor (1 GHz Pentium recommended)

- 192 MB RAM (256 MB or more recommended)

- Video (800 x 600 or higher resolution) monitor with at least 256 colors (1024 x 768 High Color 16-bit recommended)

- CD-ROM or DVD-ROM drive

- Microsoft mouse or compatible pointing device

Sample Code

This book makes extensive use of sample code, and I know that my readers like to run and observe and modify as part of their learning process. Rather than bind a CD-ROM into the book, which is expensive and hard to upgrade, I've made a web site for this book and placed the sample code on it. You'll find it online at *www.programcab.com* or *www.programscsf.com*. In addition to the sample code from this book, I'll also be placing new articles of interest and code samples on it. So check it out, try the code, and tell me how you like it. All of the sample code is written with the May 2007 version of SCSF.

Acknowledgments

This book, as with any other, is the product of many minds. I'd like to first and foremost thank the CAB and SCSF development team, which started with Eugenio Pace, Peter Provost, and Steve Elston. New team members such as Blaine Wastell and Glen Block gave invaluable advice. On the Microsoft Press side, Ben Ryan did his customary great job as acquisitions editor, as he's done for me going on a decade now. Devon Musgrave and Valerie Woolley helped put the plans into action. As always, the greatest thanks go to all of my students. Teaching something is how you learn it. By inviting and allowing me to teach them, challenging me with good questions, and constantly asking, "Yeah, but what if ..." , they've forced me to delve into and take apart this topic. I wouldn't have gotten anywhere near as far on my own. As I always say at the end of a training class, "Thank you all for coming, because it would suck doing this all by myself."

Support for This Book

Every effort has been made to ensure the accuracy of this book and the companion content. As corrections or changes are collected, they will be added to a Microsoft Knowledge Base article. Microsoft Press provides support for books and companion content at the following Web site:

http://www.microsoft.com/learning/support/books/

Questions and Comments

If you have comments, questions, or ideas regarding the book or the companion content, or questions that are not answered by visiting the sites above, please send them to Microsoft Press via e-mail to

mspinput@microsoft.com

Or via postal mail to

Microsoft Press
Attn: *Programming Microsoft Composite UI Application Block and Smart Client Factory* Editor
One Microsoft Way
Redmond, WA 98052-6399
Please note that Microsoft software product support is not offered through the above addresses.

Chapter 1
Introduction

A.   Problem Background

1. The superiority of rich client applications over browser-based applications for dedicated users has been proven time and again. I once did a study for a health care client in which I calculated that every extra mouse click in a user interface, when multiplied by the number of PCs in the enterprise and the frequency of their usage, cost the client $125,000 per year in extra employee time. Every. Single. Click. (Bang! just went a third of a million dollars in those last three words.) So optimizing the user interface is not optional for an industrial-strength line-of-business application.

    Even more important than lowering this constant friction, a better user interface lowers the chance of a catastrophic error. Consider the case of a hospital system, where a doctor wants to view information about a specific patient. An individual patient's data is probably scattered in separate silos—the pharmacy system, the X-ray system, the surgery scheduling system, etc. Each of these has its own data, its own storage and retrieval system, its own user application, its own jealously guarded fiefdom. The doctor has to use a separate application to access each silo, according to each application's rules, with little or no co-ordination among them, using his own brain to string together these separate pieces of information into his own mental context describing one specific patient. At best, this system consumes time and mental energy that the doctor would rather use for thinking about what's wrong with the patient and how to fix it. At worst, it leads to catastrophic errors, as you can see about to happen in Figure 1-1. Possibly a cut or paste operation failed and the doctor didn't notice. ("Oops. Dang. Sorry. Do you want us to try to sew that leg back on?")

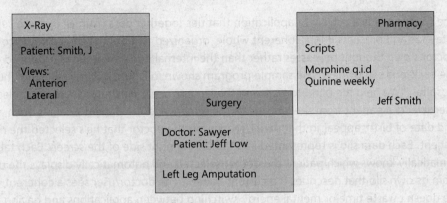

FIGURE 1-1 Separate hospital applications.

2. The doctor needs a rich client application that ties together data from all of these separate silos and presents it in a coherent whole, organized and optimized according to the doctor's own thought processes rather than the internal implementations produced by the back-end silo owners. The sample program shown in Figure 1-2, from Rolling Thunder Hospital ("We help the blind to walk and the lame to see"), illustrates these principles. The user selects a patient from the left side of the screen, and the patient's details, such as sex and date of birth, appear in that panel, reassuring the doctor that he's selected the right patient. Each data silo is represented in a tab on the right side of the screen. Each tab automatically knows which patient the doctor selected and automatically displays the data from its own silo that describes the current patient. The doctor/user sees a coherent whole. He doesn't waste time or mental energy switching between applications and having to memorize the behavior, often contradictory, of all the separate applications. And the sort of error that he made in the previous example is much harder (not impossible, nothing's impossible that involves human error, but hardER) to make than it was before. The world is a better place. That's what we need.

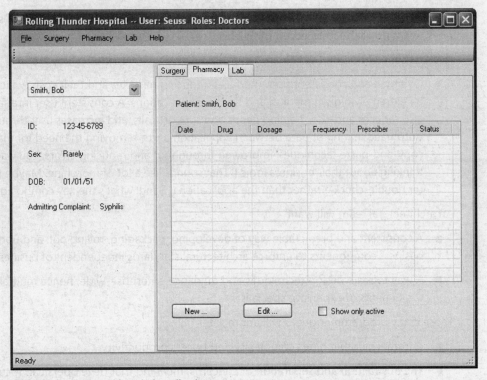

FIGURE 1-2 Integrated hospital application.

3. OK, that's the sort of application that we need to produce. From a software design standpoint, what do we need to accomplish it?

From a user standpoint, the requirements are relatively easy to state. They are

- Common and consistent user interfaces, reducing mistakes and their costs, which can be extremely high in medical and financial applications. A consistent user interface also reduces end-user training needs and support calls, and increases user throughput. Automatic sharing of data between application parts, removing the need for manual copy-and-paste techniques, improving throughput and reducing errors. You want users thinking about their business logic ("That sounds like a lot of morphine. Maybe I'd better double-check") rather than the application ("Dang, what's the key combo again?").

An architectural team will want

- A consistent and predictable way of developing, packaging, rolling out, and updating business components to enforce architectural standards independent of business logic.

- A way to scale design techniques and guidance enterprise-wide, hence multiplying the benefits of expensive specialists.

Those crazy programmer geeks will appreciate

- A framework that hides complexity and boosts productivity.

- The abstraction and separation of concerns, meaning that developers can focus solely on business logic or user interface (UI) design or infrastructure services without requiring knowledge of other parts of the overall application.

Operations teams, the people charged with deploying, administering, and supporting applications, often feel that they get the short end. To some extent, this feeling is justified. Programmers don't like thinking about that kind of stuff because to them it isn't fun. But as the combat infantryman always says, "You're just as dead." Money spent on operations staff is just as spent; errors and downtime caused by operations issues are just as erroneous and just as down as those caused by programming bugs. Like losing weight or getting exercise or having money left in your wallet at the end of the month, operational efficiency is a very good thing, and it doesn't just happen on its own. You have to go out and make it happen; otherwise, it won't.

Operations teams would like

- The consolidation of shell applications resulting in the need for only one executable file to be shipped, reducing the number of potential common language runtime (CLR) versioning issues.

- Easier rollout of common business elements and modules, resulting in consistent configuration management and instrumentation implementations across a suite of applications.

- A pluggable architecture that enables basic services (such as authentication and catalog provisioning) to be driven from server-side infrastructures, which, in turn, enables central management of many smart client applications at a time.

4. Now that we see what the application needs, what sort of design would best produce it? There's an awful lot of stuff going on inside it. The main screen contains multiple collaborating parts, each one addressing a specific aspect of a business process: patient selection, X-rays, surgery scheduling, and so on. We need the parts to be integrated visually to provide a consistent user experience, and we need to share information with each other to some extent.

Ten years ago, the classic design would have been to build a monolithic application in which every part knew about every other part at compile time. But that design doesn't work well in today's enterprise, because the difficulty of developmentincreases exponentially as program size increases. The application quickly gets so complicated that no one can keep track of all the side effects and back channels, and it becomes unmaintainable, unextendable, and unusable. Microsoft Office is the classic example of a complex monolithic application.

Instead, it would be better to separate the parts of the program and reduce their dependencies on each other to a minimum. They'd be developed by different teams. Since each part knows less about the other parts, the side effects would be minimized, and the development effort would approach linearity to size of the program. The parts would interact independently with their own back-end systems and could be independently versioned, deployed, and updated.

Consider the evolution of program design in the past 30 or so years. We went from *GoTo* statements to functions and then from functions to objects. With each step, we knew less about each piece of the program so that we could effectively deal with more pieces. We've gone about as far as we can go with tightly coupled objects, and we now need to take the next step.

B.   Solution Architecture: Loose Coupling with CAB

1. We can solve these architectural problems by developing an application from loosely coupled parts. Instead of compiling everything together into one giant .exe file, we can build the parts more or less separately and stitch them together ("Compose them") at runtime using services provided by the Composite UI Application Block (CAB).

   This approach allows our application to be based on the concept of modules or plug-ins. Because of loose coupling, there are fewer interactions between the parts, making our application not only easier to develop initially, but also easier to extend and maintain than a classic monolithic application. It also allows development teams to work independently—for example, the user interface specialists concentrating on presentation and the business logic teams on business logic, as shown in Figure 1-3.

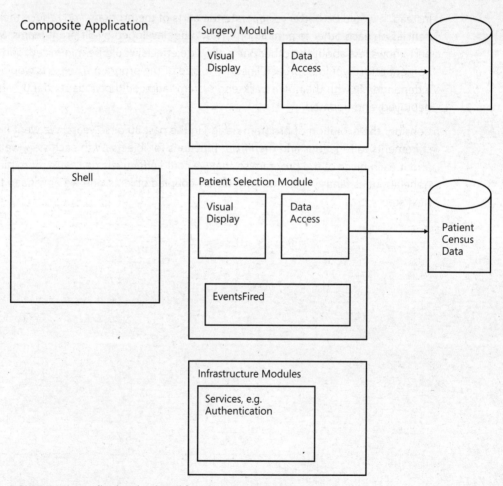

FIGURE 1-3 An application developed from loosely coupled parts.

2. We've seen this sort of architecture before in any number of modular designs. The key improvement that CAB provides is **a prefabricated infrastructure that supports this loose coupling**. For example, instead of having to know at compile time which class is responsible for firing an event (announcing, say, the user has selected a patient from a list or a search), CAB allows us to easily write code that says, "Whenever the event with the name *PatientSelected* is fired, no matter who does it, call this response function to tell me about it because I care." Or "Whichever object is in charge of implementing the interface *IWhatever*, give me a reference to it because I need it to get my work done." Or "Take this control and show it in the display windows whose name is [this] so the user can see it, please." CAB makes extensive internal use of .NET Reflection to provide these capabilities. The services that CAB provides for accomplishing these tasks are listed in Table 1-1, along with the lesson in which each is described.

TABLE 1-1 Location in this Book of CAB Functionality for Solving Loose Problems of Loosely Coupled Applictions.

Problem	CAB Infrastructure	Chapter
Central control of modules loaded at runtime by a specific user	Module enumerator and loader services	2
Connection to programmatic services	Service infrastructure	2
Connection of logical subapplications to each other within a composite application	*WorkItem* mechanism and dependency injection	3
Display of visual controls created by subapplications	Workspaces (container frames) and SmartParts (controls container with those frames)	4
Different subapplications sharing main menu, toolbar, and status bar	User interface extension sites	5
Notifying and being notified of events	Publish and subscribe event broker service	6

3. Table 1-2 identifies the main terms we will be using in our discussion of CAB, listed in the order in which you're likely to think about them. A sample CAB application is shown in Figure 1-4, illustrating some of the concepts.

TABLE 1-2 Definitions of Basic Terms Used in a CAB Application

Shell Application	The main Windows Forms application, the outer container of all parts of a CAB application.. The Shell Application manages the CAB startup process.
Shell Form	The main window of the Shell Application. The generic term *Shell* used on its own usually means *Shell Form*, although not always. It usually contains *Workspaces* and user interface elements such as menus and toolbars
Workspace	The container window for a SmartPart owned by a *WorkItem*;. The *Workspace* can control how the SmartPart is displayed or hidden. The CAB provides several standard *Workspace* classes, and you can also write your own.
WorkItem	A runtime container of the objects and services used by a discrete part of a CAB application. Think of it as a logical sub-process or sub-application. It is the basic unit of software scoping in a CAB application. Your business logic lives in one or more *WorkItems*.
SmartPart	A visual presentation, a view, of the data owned by a *WorkItem*; it is owned by a *WorkItem* and displayed in a *Workspace*. A SmartPart is usually implemented as a Windows Forms User Control, often containing other Windows Forms controls. Besides displaying the data of a *WorkItem*, it often allows the user to modify it.
Service	A supporting class that provides programmatic functionality to other objects in a loosely coupled way. It usually contains utility methods that are not tied to a specific work item.
Module	A .NET assembly that provides the physical container for *WorkItems*, services, and their supporting classes.

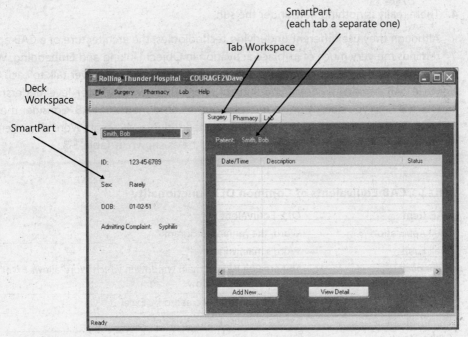

FIGURE 1-4 Application diagram of shell, smartparts, and workspaces.

4. There really is nothing new under the sun.

Although they use different underlying technologies, the architecture of a CAB application reminds me very much of in-place activation in Object Linking and Embedding. While not a perfect or complete analogy (OLE-embedded objects rarely if ever talk to each other, while CAB items often do), if you think of things that way, your user-level understanding of OLE will help you quickly grasp the roles of the components of CAB. Consider the archetypal case of an Excel spreadsheet embedded in a Microsoft Word document as shown in Figure 1-5. The CAB analogy would be as shown in Table 1-3.

TABLE 1-3 CAB Equivalents of Common OLE Functionality

CAB Item	OLE Equivalent
Shell Application	Word, the running program
Shell Form	Word's main window
Workspace	The area on Word's main window in which Word allows Excel to display its editing window
SmartPart	The editing window created by Excel
WorkItem	Excel's business logic
Service	OLE libraries
Module	The physical file containing the business logic code, Excel.exe

Shell Form (main program's
main window)

Workspace (main program's frame for
display of subordinate visual parts)

SmartPart (subordinate
visual display)

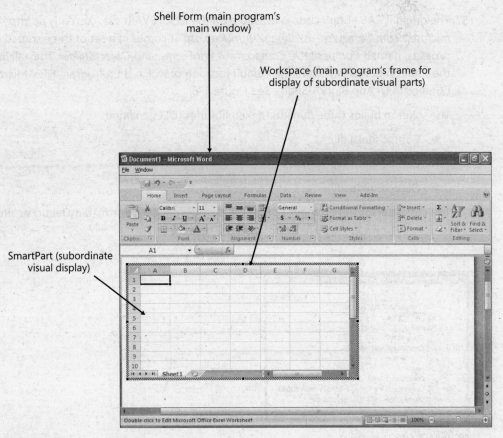

FIGURE 1-5

5. The original CAB library code is available on the MSDN Web site, currently at *http://msdn. microsoft.com/library/en-us/dnpag2/html/cab.asp*. It comes in a set of three source code projects, named *CompositeUI*, *CompositeUI.WinForms*, and *ObjectBuilder*. They all live in the solution *CompositeUI.sln*, the default location of which is C:\Program Files\Microsoft Composite UI App Block\CSharp. See Figure 1-6.

The solution builds three dynamic linked libraries (DLLs), named

- CompositeUI.dll

- CompositeUI.WinForms.dll

- ObjectBuilder.dll

The easiest place to find them all is in the CompositeUI.WinForms\bin\Debug or \Release folder, as shown in Figure 1-7.

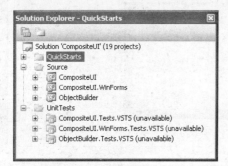

FIGURE 1-6 CAB DLL projects.

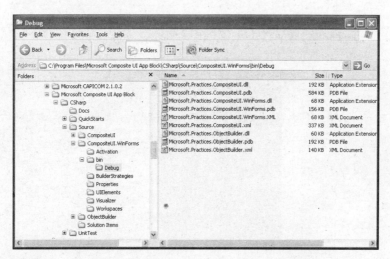

FIGURE 1-7 CAB DLLs built in a project folder.

6. You will have to decide early on in your project exactly how you want to distribute these DLLs. If you make every CAB application require its own private copy of them, then you can bundle the whole thing up and download it via *ClickOnce*. This is fast and easy and hard to get wrong. The only disadvantage is that you will then have extra copies of them. They aren't very large, about 300KB total, so perhaps this isn't too bad.

   Alternatively, you could sign them with your strong name and place them in the Global Assembly Cache GAC. However, this means that *ClickOnce* won't be able to put them there because it can't modify the GAC.

C.   Smart Client Software Factory (SCSF)

1. Feedback from the original release of CAB provided an object lesson to Microsoft. Developers liked the idea of a prefabricated toolkit for writing rich client applications. However, these developers reported that they found CAB difficult to use for two main reasons.

First, few tools existed for helping them develop CAB applications. There were, for example, no wizards to generate the shell of a CAB application. Developers had to generate a standard Windows Forms project and modify the application files to make it use the CAB libraries. They had to cut and paste lots of code, do more repetitive grunt work than they had to do with other types of applications—more than they felt that they should have had to. If you remember writing an original Windows Software Developer Kit (SDK) program (*RegisterClass*, *CreateWindow*, *ShowWindow*, *GetMessage*, *TranslateMessage*, *DispatchMessage*, Window response function, *WM_PAINT*, etc.), beginning every project with an extensive cut-and-paste session from some other Windows project, you'll understand the feeling. The architectural goodness of CAB was apparent to developers, but many felt that the price of grunt work necessary to use it was higher than it should have been, higher than they were willing to pay. For another analogy, think of the Microsoft Foundation Classes (MFC): nice prefabricated toolkit. Now imagine it without Class Wizard, its main automation tool. You could use it and maybe still make a profit over using the SDK, but too much of the time savings from prefabrication would get lost in the cutting and pasting. That's what CAB was like on its first release.

Second, since CAB was very new, a number of architectural issues hadn't been worked through completely by the time of its release. Here's one small but omnipresent example: How should the various work items share and communicate with the status bar's main message string? Expose it directly as a class? Place it on the root work item? Do something else? Here's another: Should you derive a new class for each *WorkItem* or use the same *WorkItem* class for each one and express differences in business logic through some sort of object contained within the *WorkItem*? Developers still had architectural work to do on the infrastructure of their rich client programs. They hadn't yet been completely freed to concentrate only on their business logic.

2. To solve these problems, Microsoft release the Smart Client Software Factory, or SCSF (originally called SC-BAT). This is a set of code, documentation, and tools that illustrate and ease the task of following the design patterns that experience has shown are optimal for most CAB applications. The first version shipped in July 2006, and the second version in May of 2007, just before this book went to press. SCSF contains three main enhancements to baseline CAB.

First, SCSF contains a number of classes that provide brand-new functionality that CAB developers wanted but that weren't included in the base release. For example, the 2006 release contains the *Dependent Module Loader* service and the *ActionCatalogService*. The 2007 release contains support for Windows Presentation Framework , as well as classes for handling disconnected and intermittently connected operation. This functionality is included in the SCSF and not in the baseline CAB assemblies. If you think of these additions as CAB .1 and .2, which get added to the baseline CAB libraries to produce versions 1.1 and 1.2, you'll have the right mental model.

Second, SCSF contains classes that provide architectural guidance to developers. For example, the *ControlledWorkItem* class (see Chapter 3) is sealed, thereby indicating to the developers who use it that they should not derive new classes from it. It also provides easy access to and from a *Controller* object, thereby indicating to developers that business logic belongs on this controller.

Third, SCSF contains a set of tools known as a *Guidance Package*. This Visual Studio plug-in generates code that implements the recommended design patterns. If you think of this as analogous to the MFC Class Wizard, you'll have the right mental model.

3. Platt's Third Law of the Universe states simply: "Laziness trumps everything." According to this law, all human beings are inherently lazy and will do the absolute minimum amount of work they can get away with under any circumstances. Therefore (first corollary), that which is easy to do will be done frequently, whether it should be or not, and that which is hard to do will be done infrequently, whether it should be or not. Therefore (second corollary), a good design makes the good, safe, smart things easy to do, and the bad, dangerous, stupid things hard to do. SCSF makes writing good code easier than writing bad code (mostly, usually); therefore, more good code will be written and less bad code.

The first generation of CAB developers, whom I taught in the first half of 2006, were drawn to CAB by its modular architecture, especially for large and complex applications. They felt that following the CAB design patterns would be good for their applications in the long run even if it took more effort today, like flossing their teeth. With the release of SCSF, these developers are happier because the flossing got easier, an unexpected advantage. But the release of SCSF also attracted an entirely new set of developers who view SCSF-CAB as a convenient toolkit for producing even small applications more quickly and easily than they previously could. They see an economic gain from prefabrication, again, conceptually similar to the way the MFC and its wizards made desktop application development easier than a raw SDK application. To take the dental analogy one step further, they perceive that this new automatic brusher-flosser machine (SCSF-CAB) is faster and easier than brushing alone the old manual way (plain Windows Forms).

Clients often ask me which version of CAB to use, baseline or SCSF? I always recommend the SCSF version. Not only is it much easier to use, but it contains the invaluable experience of the first generation of CAB developers. I will use the SCSF version of CAB throughout this book, except in the next section of this chapter.

4. The May 2007 release of SCSF installs signed binary versions of CAB DLLs. You do not have to install the baseline release of CAB to use it. The CAB library source code is automatically installed when you install the SCSF source code. The course installs itself in your Visual Studio 2005 Projects folder.

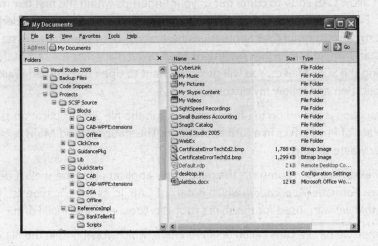

D.   Simplest Application Walkthrough, Classic CAB

1. The Walkthrough QuickStart application, which comes with the baseline CAB download and also the May 2007 SCSF, is most developers' first introduction to CAB. Its step-by-step instructions allow CAB code to come out of your fingertips within the first few minutes of touching it. Even though some of its architectural choices have been superseded, I'm going to start my discussion of CAB applications here, so you can see where we started from and understand why these further evolutions have taken place. For the rest of this book, I will be using the SCSF dialect. You will probably want to open the finished Walkthrough sample application and follow my discussion.

    The *Main* function is shown on the facing page. It's in the file ShellApplication.cs, which takes the place of Program.cs in a CAB application. The static method **Main** tells the program loader where to start.

    The class *ShellApplication* represents the main outer application of the shell. It derives from the base class *FormShellApplication*, which is part of CAB. In its generic type list, we pass the classes that we want used for the shell's root *WorkItem* and the Shell Form.

    For the former, we have a class called *ShellWorkItem*, which you'll find in the file ShellWorkItem.cs. In this application, this class doesn't do anything at all other than sit there and exist as the CAB application's root *WorkItem*. This pattern of deriving additional classes from *WorkItem* is not used today, as I discuss in Chapter 3. Today, we use plain old *WorkItem* for the class of the root *WorkItem*. But it was prevalent for development of CAB and the first six months or so of its released lifetime, so you will see it in many older applications and documentation.

    The Shell Form is the main window that the shell application creates. This is the place where you put your workspaces and user interface items.

    When the application starts running, the CAB framework creates an instance of the root work item and an instance of the Shell Form. Then it loads the modules called out in the ProgramCatalog.xml configuration file, which is the next thing we'll look at (see Figure 1-8).

```
using System;
using System.Collections.Generic;
using System.Text;
using Microsoft.Practices.CompositeUI.WinForms;

namespace ShellApplication
{

    // This class is that base of the application. You pass it
    // the types of the root WorkItem and the root Windows Form.

    public class ShellApplication : FormShellApplication<ShellWorkItem,
        ShellForm>
    {

        // Here's the program's entry point

        [STAThread]
        static void Main()
        {
            // Create an instance of the application class and call
            // its Run method to start it

            new ShellApplication().Run();
        }
    }
}
```

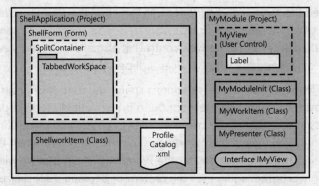

FIGURE 1-8 Architectural diagram of CAB Walkthrough sample.

2. During program startup, the shell application loads the modules that are specified in the startup file ProfileCatalog.xml. Each *<ModuleInfo>* element specifies a module file to load into the CAB application. The file looks like this:

```xml
<?xml version="1.0" encoding="utf-8" ?>
<SolutionProfile xmlns="http://schemas.microsoft.com/pag/cab-profile" >
   <Modules>
        <ModuleInfo AssemblyFile="MyModule.dll" />
   </Modules>
</SolutionProfile>
```

When a module is loaded into the shell application's address space, the loader looks for classes that derive from the CAB base class *ModuleInit*. Such a class is shown on the facing page. This is the place where you put initialization code.

The first thing you see when you look at the class is a property called *ParentWorkItem*, which is of class *WorkItem*. It has the curious property of being write-only; that is, it can be set from outside but not read. It is marked with a *[ServiceDependency]* attribute. This causes a *dependency injection*, an important part of the loose coupling strategy of CAB. The module loader in CAB sees the attribute and knows that it means that the module needs the specified type—in this case, *WorkItem*. The term *Service* is ambiguous here. It's used in the CAB meaning of "any class accessed through this loose coupling mechanism," not specifically the *Services* collection discussed in Chapter 2. So when the module is loaded, the CAB framework sees that it depends on the *WorkItem* type, which means the *WorkItem* that caused the module to be loaded—in this case, the root *WorkItem*. It therefore sets the property with a reference to that work item. This reminds me somewhat of the old *IConnectionPoint* interface in ActiveX controls. It also allows a type viewer application to examine the object and see what services it depends on.

Once the module is loaded in the process's address space, the framework calls its Load method, which we override to make a place for our module's initialization code. First, we call the base class as usual. Next, we create a child work item and place it into the *WorkItems* collection (which means child *WorkItems*) of the parent *WorkItem* that we got from the injection. This technique creates the *WorkItems* chain that we will see in Chapter 3. This sample program again uses the pattern of deriving a specialized class from *WorkItem*—in this case, *MyWorkItem*. As we will see in Chapter 3, this pattern is no longer used.

In the sample application, the child *WorkItem* wants to display a SmartPart, so it needs to know the workspace that it's supposed to use for this purpose. In this example, our initialization code fetches the workspace from the parent *WorkItem* by means of a string name (again, loose coupling) and passes it into the child *WorkItem* in its Run method. This method is not a standard part of the *WorkItem*; instead, it's been added for this purpose. We don't usually follow this pattern any more. As we'll see in Chapter 4, the child *WorkItems* will generally query for their workspaces directly.

```
public class MyModuleInit: ModuleInit
{
    private WorkItem parentWorkItem;

    // This write-only property is marked as being dependent on the
    // ParentWorkItem service. This causes the CAB framework to inject
    // that into the property at creation time.

    [ServiceDependency]
    public WorkItem ParentWorkItem
    {
        set { parentWorkItem = value; }
    }

    // We override the Load method to perform the module's
    // initialization

    public override void Load()
    {
        // Call the base class

        base.Load();

        // Create a new child work item and place it into the parent
        // WorkItems collection.

         MyWorkItem myWorkItem =
                parentWorkItem.WorkItems.AddNew<MyWorkItem>();

            // Call the new work item's Run method, thus starting it up.
        // In this design, the Run method requires the workspace in which
        // it shows its user interface.

        myWorkItem.Run(parentWorkItem.Workspaces["tabWorkspace1"]);
    }
}
```

3. Our *WorkItem* code is shown on the facing page. The *ModuleInit* calls the Run method of the *WorkItem*, passing it a *Workspace*, which is a little confusing. The base class has a Run method, which fires an event and calls *OnRunStarted*, but it doesn't take a parameter as this one does. This sample has overloaded the Run( ) method to accept an *IWorkspace* interface. Most of the other examples use a Show( ) method. The modern way of doing this is to place all of this logic into a *Controller* object, associated with the *WorkItem* at creation time, as discussed in Chapter 3.

This example uses the Module – View – Presenter (MVP) architecture, which I discuss in section 4. For now, look at the Run method and note that it creates a thing called a *view* and a thing called a *presenter*. Note that, after (or sometimes during) the creation process, it adds these new things to its own *Items* collection. There's no reason that the *WorkItem* couldn't retain references to these objects in member variables, and some of them indeed will do so. However, adding them to the *Items* collection is the first time the CAB framework gets a look at them—for example, to place them in other collections or to perform dependency injection. To work with CAB, they need to first be shown to CAB, as baby Mowgli had to be shown to the Seeonee wolf pack in Rudyard Kipling's *Jungle Book*. ("Look well, O Geeks.")

The *WorkItem* code looks like this:

```
public class MyWorkItem: WorkItem
{
    // This method is called at startup time by
    // the ModuleInit code. Production applications tend
    // to call this method Show( ) instead.

    public void Run(IWorkspace tabWorkspace)
    {
        // Create the new view control class that we use to display this
        // information. Add it to our Items collection

        IMyView view = this.Items.AddNew<MyView>();

        // Create the presenter object that will manipulate the view
        // Add it to out Items collection as well.

        MyPresenter presenter = new MyPresenter(view);
        this.Items.Add(presenter);

        // Tell the workspace to show that view

        tabWorkspace.Show(view);

    }
}
```

4. This sample uses the Model – View – Presenter design pattern, shown in the diagram in Figure 1-9. You can find it discussed in excruciating detail by Martin Fowler online at *http://www.martinfowler.com/eaaDev/ModelViewPresenter.html.*

The basic idea is to implement a three-tier architecture in the client application. The *WorkItem* is the model referred to in the diagram. It contains the program's data, the state with which the user is concerned. The business logic lives in the presenter, here represented by the *MyPresenter* class we saw being created on the preceding pages. This class takes the model's state, thinks about it, and modifies the actual presentation done by the view. The view is a new object of class *MyView*. It is a Windows Forms control that is a SmartPart. It contains the actual controls that display data to the user.

We first have to define an interface to govern the communication between the presenter and the view. In this example, it's the *IMyView* interface shown at the top of the facing page. It contains one property of type string, called *Message*, and one event, called *Load*. The presenter doesn't need or want to know how these are implemented in the view.

If you didn't have this separate layer of the presenter, you would have the model interacting directly with the view. That is not an evil thing in and of itself. But this means that the view and the model are directly coupled, which makes it harder to develop them.

The presenter's code is shown on the facing page. At the time of its construction, it is given by the work item that constructs the view that it is to govern. It saves a reference to this view and hooks up an event handler for the view's *Load* event. In that handler code, it sets the view's *Message* property when it receives the event.

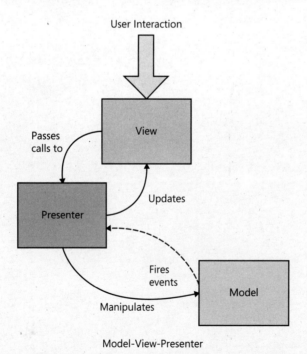

FIGURE 1-9 Model-View-Presenter patterns used in a CAB application.

First, we define the interface to be used between the presenter and its views. In the case of the sample, it looks like this:

```
public interface IMyView
{
    event EventHandler Load;
    string Message { get; set; }
}
```

The presenter code looks like this:

```
public class MyPresenter
{

    IMyView view;

    // When this presenter is constructed, we get passed
    // the view on which we are to display

    public MyPresenter(IMyView view)
    {
        // Save a reference for future use

        this.view = view;

        // Hook up a handler to the event fired by the view

        view.Load += new EventHandler(view_Load);
    }

    // The control fired this event. Respond to it.

    void view_Load(object sender, EventArgs e)
    {
        view.Message = "Hello World from a Module";
    }
}
```

5. Finally, we reach the view itself. The view is meant to be a pure presentation object. It represents one particular way of looking at (and possibly modifying) data. If you remember the document-view architecture from the MFC, you'll have a decent mental model here. As an application designer might choose to present several different views of a document (a zoomed-out page print view, a zoomed-in details view, etc.), so an application designer here may choose to present views in any way that she thinks will make the user's life easier.

The view's code is shown on the facing page. Note that it derives from the Windows Forms base class *UserControl* and implements the interface *IMyView*. You can see the *Message* property implemented in code beneath it. The *Load* event is not shown in the code. The *UserControl* base class's *Load* event is considered the implementation of this interface member.

Looking at the property *Message*, you can see that it sets the property of a label that the *UserControl* contains. The hiding of this implementation is, to my mind, the main benefit of the MVP architecture. The presenter doesn't know or care how this feature is implemented. If the user interface design requires moving things around such that the message appears on Label 2 instead of Label 1, that doesn't affect the presenter in any way.

The key part of this design from a CAB standpoint is the presence of the *[SmartPart]* attribute. It tells CAB to treat it as SmartPart and thus to manipulate it in accordance with the commands of the workspace in which it appears.

```
[SmartPart]
public partial class MyView : UserControl, IMyView
{
    public MyView()
    {
        InitializeComponent();
    }


    public string Message
    {
        get
        {
            return this.label1.Text;
        }
        set
        {
            this.label1.Text = value;
        }
    }

}
```

E.   Tracing and Visualization

1. One of the most important things that anyone can learn about a new piece of software is how to pick it apart and see what it is really doing. There are two main methods for doing that with CAB. The first, shown next, is with classic debugger tracing. The CAB libraries have been written to output debugger traces in strategic points. All you have to do is turn it on by making the entries shown below in your app.config file. The source name is the name of the class, such as the *ModuleLoaderService* shown here:

```
<system.diagnostics>

 <switches>
  <add name="MySwitch1" value="all" />
 </switches>

  <sources>
        <source
             name="Microsoft.Practices.CompositeUI.Services.ModuleLoaderService"
                          switchName ="MySwitch1"/>
  </sources>

</system.diagnostics>
```

The tracing shows up in the debugger's output window, as shown in Figure 1-10.

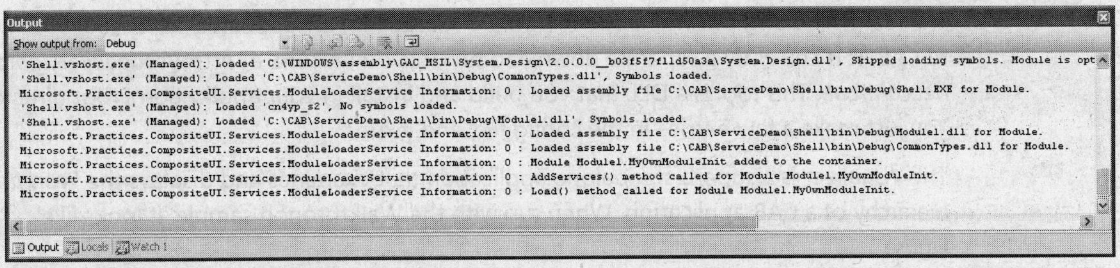

Output

Show output from: Debug

```
'Shell.vshost.exe' (Managed): Loaded 'C:\WINDOWS\assembly\GAC_MSIL\System.Design\2.0.0.0__b03f5f7f11d50a3a\System.Design.dll', Skipped loading symbols. Module is opt
'Shell.vshost.exe' (Managed): Loaded 'C:\CAB\ServiceDemo\Shell\bin\Debug\CommonTypes.dll', Symbols loaded.
Microsoft.Practices.CompositeUI.Services.ModuleLoaderService Information: 0 : Loaded assembly file C:\CAB\ServiceDemo\Shell\bin\Debug\Shell.EXE for Module.
'Shell.vshost.exe' (Managed): Loaded 'cn4yp_s2', No symbols loaded.
'Shell.vshost.exe' (Managed): Loaded 'C:\CAB\ServiceDemo\Shell\bin\Debug\Module1.dll', Symbols loaded.
Microsoft.Practices.CompositeUI.Services.ModuleLoaderService Information: 0 : Loaded assembly file C:\CAB\ServiceDemo\Shell\bin\Debug\Module1.dll for Module.
Microsoft.Practices.CompositeUI.Services.ModuleLoaderService Information: 0 : Loaded assembly file C:\CAB\ServiceDemo\Shell\bin\Debug\CommonTypes.dll for Module.
Microsoft.Practices.CompositeUI.Services.ModuleLoaderService Information: 0 : Module Module1.MyOwnModuleInit added to the container.
Microsoft.Practices.CompositeUI.Services.ModuleLoaderService Information: 0 : AddServices() method called for Module Module1.MyOwnModuleInit.
Microsoft.Practices.CompositeUI.Services.ModuleLoaderService Information: 0 : Load() method called for Module Module1.MyOwnModuleInit.
```

Output　Locals　Watch 1

FIGURE 1-10 CAB code outputting traces in debug window.

2. The real whiz-bang tracing happens with a Visualizer, an unsupported utility tool specific to CAB. It is currently a separate download on GotDotNet.com, at *http://codegallery.gotdotnet.com/cab*. This tool is a DLL that you build and install in your application's root directory. You just make a few entries in the config file, and it runs.

   The Visualizer is a superb standalone application that examines and displays the *WorkItem* hierarchy of a CAB application. When run with the Walkthrough sample, it looks like Figure 1-11.

   In the figure, the Visualizer shows the root work item, in this case of class *ShellWorkItem*. Its properties appear in the window on the right side.

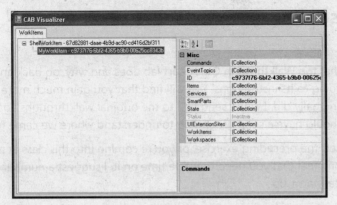

FIGURE 1-11 Sample visualizer program provided with CAB.

Chapter 1 Lab Exercise
Introduction

1. Now that you know what the Walkthrough QuickStart lab does and why, go back and run through it again according to its instructions. You will find that you gain much more benefit from it than you originally did. I won't send you to the original walkthroughs again, but I do believe that you should have a solid grounding to understand where we came from.

2. If you've already finished the preceding exercise, or you're coming into this class at a more advanced level and don't feel that you need to waste time on it, I suggest a number of alternatives for a mental warm-up:

    a) Read the following lessons and prepare questions, things that have always bothered you about CAB and you want answered soon;

    b) Try one of the more advanced labs in a later chapter, such as the web service for specifying user interface elements that you'll find at the end of Chapter 5;

    c) Open the reference applications, such as the bank branch or appraiser workbench, and step through them from the beginning;

    d) Try the lab for the lesson on Generics, which comes after Chapter 7. A quick brush-up on generics is a good idea because CAB and SCSF use them extensively; or

    e) Use the lab time to clear out your email inbox and return all your messages, thus clearing off your plate for the more challenging assignments ahead.

Chapter 2
The Shell and Services

A.   Concepts and Definitions

1. The *Shell Application* is the .NET application that provides the outermost container of your CAB application, containing the Main() method that starts a program's run. It is generally stored in its own project and written by the highest-level application development team.

   The *Shell Form* is the main Windows Forms window of this application.

   The generic term *Shell*, used on its own, sometimes means *Shell Form* (as in the method *AfterShellCreated*), sometimes *Shell Application* (as in the method *AfterShellCreated* in the Windows Presentation Framework (WPF) implementation discussed in Chapter 8), and sometimes the combination of the two together.

   As we saw in the first chapter, a CAB application contains a hierarchical tree of *WorkItem* objects. The root *WorkItem* of this tree lives in the shell application. It is called the *Root WorkItem* or, less commonly, the *Shell WorkItem*.

   A *Service*, that so-often-overloaded term, has a specific meaning in terms of CAB. In this case, it means a singleton object of any class that is available through the loosely coupled connection mechanism of CAB, which we see in the second half of this chapter.

B. Generating a CAB Project with SCSF

1. The SCSF contains a wizard for generating the projects that comprise a CAB application. After installing the SCSF and its associated utilities, you see the Guidance Packages item in the New Project dialog box in Visual Studio. The Smart Client Development item underneath it contains a template named Smart Client Application, as shown in Figure 2-1.

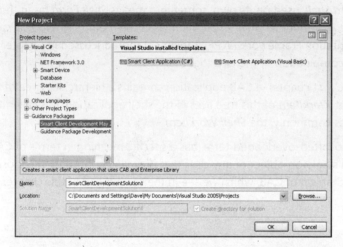

FIGURE 2-1 Visual Studio template for generating Smart Client Application.

2. The SCSF needs to know the location of the CAB dynamic linked libraries (DLLs) and of the Enterprise Library DLLs because it uses both in its generated code. It prompts you for them using the box shown in Figure 2-2. You also need to enter the namespace to be used for the classes in your generated project.

> **Note** Not all SCSF project developers like using the Enterprise Library in their CAB programs. You can actually remove it quite easily from the code generated by the SCSF if you want. Just open each generated project and remove the references to the Enterprise Library DLLs. Then compile the code and see where the errors are. There are only about three locations in which you'll have to remove *EntLib* code, generally relating to exception handling. Just make sure you put in your own exception handling in its place and don't just leave it undone.

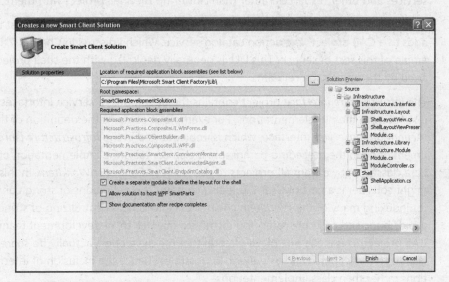

FIGURE 2-2 Wizard choices for generating Smart Client application.

3. The SCSF generates four or five projects containing many files for a complete CAB application. Figure 2-3 shows the Visual Studio Solution Explorer displaying the results of such a generation. I will now discuss the purpose of each of the projects that you see.

The Shell project, as we've seen, is the main outer container of the application. It contains classes for the shell application and the shell form. It also contains the app.config file used by most .NET applications, and a ProfileCatalog.xml file used by the default module enumerator service, which we meet later in this chapter.

The *Infrastructure.Module* project builds a DLL that is loaded by the module loader service when the project starts up. It creates a child *WorkItem*, which it adds to the *WorkItem* chain. Most CAB applications contain several modules that can create a *WorkItem*, so one of the reasons that the wizard generates this module is to demonstrate how it's done. A production application would probably use an infrastructure module to contain its own services and other resources rather than clutter up the shell project with them.

The *Infrastructure.Library* project contains the implementation of services that the SCSF adds to a CAB project. The action catalog service, which I discuss in Chapter 7, is an example of this sort of service. This DLL is generally deployed with the shell project and managed by the infrastructure team.

The *Infrastructure.Interface* project contains the definitions of service interfaces, as distinct from their implementations. For example, it contains the definition of the *IActionCatalogService* interface, which is implemented in the *Infrastructure.Library* project I mentioned in the preceding paragraph. It also contains the implementation of certain core classes that we want all projects to use, such as *ControlledWorkItem*. In this sense, you might consider it a part of CAB 1.1. It also contains the definitions of string constants used for indexing many of the *WorkItem* chain's collections. This centralizing of string names avoids bugs and collisions later. You distribute this DLL to all development teams in the CAB project. This is the one piece of code that the developers actually do share, so you want it to be as sparse and as generic as possible—hence the inclusion of interface definitions rather than class implementations.

The *Infrastructure.Layout* project is generated when you select the Create a Separate Module to Define the Layout for the Shell check box in the SCSF Wizard. If you don't select this box, then the layout of the user interface will then live directly in the shell project. The *Infrastructure.Layout* project will not be generated, and the shell form will contain a menu, status bar, toolbar, and two workspaces. If you do select the box, then the shell form will contain only one workspace that covers its entire surface. The *Infrastructure.Layout* DLL will contain a Smart Part (also known as a *view*) called the Layout View. This view will contain all the UI elements that the shell previously had and automatically place itself into the shell's workspace at startup time. You make this choice when you need several different shell form layouts for different configurations of the program. For example, suppose that client A wants its customer selection on the left side of the shell form, but client B wants its customer selection on the top of the shell form. The loose coupling mechanism of CAB allows you to create two or more separate *Infrastructure.Layout* modules and load one or another at runtime according to configuration files. This allows you to separate layout and program implementation, with a great increase in program flexibility. If, on the other hand, you are a corporate developer building a program solely for in-house use, you might not want to take the trouble of separating these concerns.

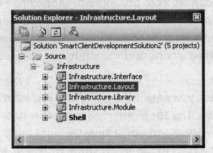

FIGURE 2-3 Projects generated by SCSF.

C.   The Shell Application and Initialization Process

1. You tend not to do a whole lot in the shell application, just as you tend not to do a whole lot in any program's main file. The shell application contains your program's entry point and performs initialization. Most of the interesting business logic lives elsewhere.

The SCSF Wizard generates a shell application class deriving from the SCSF base class **SmartClientApplication**, which in turn derives from the *FormShellApplication* of CAB. The code is shown on the facing page. You can find the former class in your app's *Infrastructure.Library* project. It adds a number of services to the base application, such as the action catalog service (*IActionCatalogService*) and the entity translator service (*IEntityTranslatorService*).

In the declaration of your application class, you pass it the types that you want the shell to use for its shell form and root *WorkItem*. The *ShellForm* class was generated for you by the wizard. If you think of it as *MyOwnShellForm*, you'll have the right mental model. The first generic parameter represents the class to use for the root *WorkItem*. In the early days of CAB, it was common to derive your own class from the base *WorkItem*, but this design philosophy has now changed, as I discuss in Chapter 3. Today, you generally use the CAB class *WorkItem* directly, as shown in the example. You start the application by calling **ShellApplication.Run( )**. To that method, we now turn our attention.

```csharp
using System;
using System.Windows.Forms;
using SimplestSCCF.Infrastructure.Library;
using Microsoft.Practices.CompositeUI;
using Microsoft.Practices.EnterpriseLibrary.ExceptionHandling;

namespace SimplestSCCF.Infrastructure.Shell
{
    /// <summary>
    /// Main application entry point class.
    /// Note that the class derives from CAB supplied base class
    /// FormShellApplication, and the main form will be ShellForm,
    /// also created by default by this solution template
    /// </summary>

    class ShellApplication : SmartClientApplication
            <WorkItem, ShellForm>
    {
        /// <summary>
        /// Application entry point.
        /// </summary>

        [STAThread]
        static void Main()
        {
#if (DEBUG)
            RunInDebugMode();
#else
            RunInReleaseMode();
#endif
        }

        private static void RunInDebugMode()
        {
            Application.SetCompatibleTextRenderingDefault(false);
            new ShellApplication().Run();
        }
    }
}
```

The startup sequence of any application is always important to understand. I show it to you in Figure 2-4. Here is what happens when you call ShellApplication.Run( ):

1. The CAB Framework creates the root work item, an object of the class passed in the generic template on the previous page (in that case, *WorkItem*)

2. Any visualizers (see the end of Chapter 1) that are specified in the app.config file are created and shown, thereby allowing a developer to monitor the internals of the CAB program.

3. CAB now adds the services that its infrastructure always needs, the list of which is hard-wired into the code. Examples of these required services are the authentication, module loader, and module enumerator services.

4. You can add new services or modify existing ones in the shell application's app.config file, as I discuss later in this chapter. These configured service operations are now performed.

5. CAB now calls the shell application's AddServices() method. If you want to add or modify any services programmatically, you override this method in your shell application class and place your code there.

6. The CAB startup code now fetches the registered implementation of its *IAuthenticationService* interface and calls its Authenticate( ) method to authenticate the user. The default implementation sets the CAB program's identity to that of the user who is logged in to the Windows desktop.

7. The main CAB program obviously has to be loaded and running for any of this to have taken place, but CAB hasn't really thought about the properties of the main program yet. The module loader service now processes the shell assembly, checking, for example, for service dependencies.

8. The root work item now has its BuildUp method called. This causes it to be scanned for dependencies, and those dependencies to be injected, being created in the process if need be. The root work item is dependent on the shell form, and this dependency causes the form to be created during this call.

9. Modules are now loaded into the CAB program. The module enumerator service fetches its list of modules, and the module loader service loads and initializes them. The default enumerator service reads its list from the ProfileCatalog.xml file. The last portion of this chapter shows you the possibility of customizing this behavior.

10. CAB calls Run on the root work item, which fires its *OnRunStarted* event.

11. CAB calls Start on its application, which in turn calls *System.Application.Run*, passing *ShellForm*. This is normal Windows Forms startup that begins to service the application's Windows message pump. This method does not return until the application shuts down.

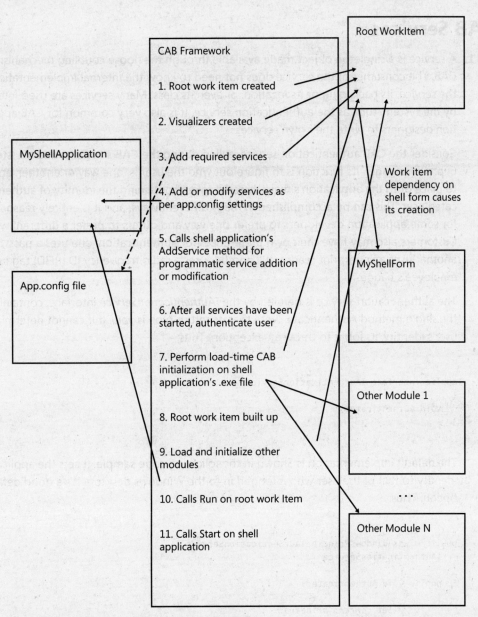

FIGURE 2-4 Startup sequence of a CAB application.

D.   CAB Services

1. A service is a singleton object made available through the loose coupling mechanism of CAB. The consumer of the service does not need to know the internal implementation of the service, its packaging or its location, or even its class. Many services are used internally by the system, such as the authentication service. It is also very common for CAB application designers to write their own services.

   Consider the CAB authentication service, called during the CAB startup sequence (step 6, previous page). Its function is to figure out who the user is, one way or another, and to place onto the application's main thread a token containing the identity of authenticated user. This can be accomplished in many different ways, and it is entirely reasonable for some application developers to prefer one way and others to prefer a different way. Customers also may have their own preferences for authentication: one uses a password, another uses a fingerprint reader, still another uses a radio frequency ID (RFID) tag in an employee's badge.

   The authentication service is defined by the *IAuthenticationService* interface, containing the single method Authenticate. Note that its return type is *void*. If it cannot determine the user's identity, its job is to throw an exception. Thus:

   ```
   public interface IAuthenticationService
   {
       void Authenticate();
   }
   ```

   The default implementation is Shown in the following code sample. It sets the application's identity to that of the user who is logged in to the Windows desktop. It's a good default option. Thus:

   ```
   public class WindowsPrincipalAuthenticationService :
       IAuthenticationService
   {
       public void Authenticate()
       {
           // Set current principal.

           AppDomain.CurrentDomain.SetPrincipalPolicy(
               PrincipalPolicy.WindowsPrincipal);
       }
   }
   ```

2. Services are placed in a collection that the *WorkItem* class maintains for this purpose (see Lesson 3 for further discussion of *WorkItems* and their collections). The collection is indexed by the interface that the service implementation supports. Only one class can be registered as the implementation of each service interface. You may place a service into this collection in a number of ways. The authentication service is initialized in the method CabApplication.AddRequiredServices, shown as item 3 in Figure 2-4. We access the root *WorkItem*, telling its *Services* collection to instantiate an object of the *WindowsPrincipalAuthenticationService* class and register it as the implementation of the *IAuthenticationService* interface. The code looks like this:

```
private void AddRequiredServices()
{
    // Tell the Services collection of the root WorkItem to instantiate
    // a new object of the specified class, and register it as the
    // implementation of the specified interface.

    rootWorkItem.Services.AddNew<WindowsPrincipalAuthenticationService,
        IAuthenticationService>();

    <… other services added >
}
```

When some other piece of code wants to fetch and use this service, it does so via the Get method of the *Services* collection of the *WorkItem*. In it, we pass the interface for which we want the current implementation. The code that is fetching the service has no idea what the implementation is or where it came from. It knows only that this is the object that is currently registered as supporting the specified interface. If the implementation has been changed from the default to a password checker or a fingerprint reader, this client code doesn't know or care—or want to know or want to care. It simply uses the registered interface to call the desired method. Thus:

```
private void AuthenticateUser()
{
    // Fetch the current object that is registered as implementing the
    // authentication service interface. We don't know or care what
    // class it is, or how it got there.

    IAuthenticationService auth =
        rootWorkItem.Services.Get<IAuthenticationService>(true);

    // Call the authenticate method. Exactly how it figures out who the
    // user is, and verifies that it really is that guy and not someone
    // else, isn't our concern.

    auth.Authenticate();
}
```

3. Now, suppose we want to replace the default authentication service in our application. Instead of just accepting the identity of the Windows desktop user, we want a different method, say a password or a fingerprint reader or an RFID tag in an employee badge. What would we do? first, we'd write a new implementation of the service, more or less like this:

```
public class MyFingerprintAuthenticationService :
    IAuthenticationService
{
    public void Authenticate()
    {

        // Perform fingerprint authentication

        MyOwnUserData user = MyOwnFingerPrintReader.GetUser ( ) ;

        // If we successfully figured out who the user was, then place
        // that information into an identity token and place the token
        // onto the main thread.

        if (user != null)
        {
            GenericIdentity identity = new GenericIdentity(user.Name);
            GenericPrincipal principal = new GenericPrincipal(identity,
                user.Roles);
            Thread.CurrentPrincipal = principal;
        }
        else
        {
            throw new AuthenticationException(
                "couldn't find your fingerprints");
        }
    }
}
```

4. Now that we've written our service implementation, how can we get it into the root work item in place of the original authentication service? There are two ways: either programmatically or declaratively. In the former case, we'd override the shell application's AddServices method, thus:

```
protected override void AddServices()
{
    base.AddServices();

    // Create an instance of our own authentication class

    MyFingerprintAuthenticationService mfas =
        new MyFingerprintAuthenticationService ();

    // Remove the default authentication

    RootWorkItem.Services.Remove< IAuthenticationService>();

    // Replace it with our own

    RootWorkItem.Services.Add< IAuthenticationService>(mfas);
}
```

Alternatively, we could make entries in the app.config file. The *<remove>* element specifies the interface for which to remove the implementation. The *<add>* element specifies the interface (*serviceType*) and the class (*instanceType*) to register in its place. Note that each type is specified by its fully qualified name, then a comma, and then the name of the DLL file in which it is located. Thus:

```
<CompositeUI>

  <services>

    <remove serviceType =
    "Microsoft.Practices.CompositeUI.Services.IAuthenticationService, CompositeUI" />

    <add serviceType =
        "Microsoft.Practices.CompositeUI.Services.IAuthenticationService,
            CompositeUI"
        instanceType="MyOwnNamespace. MyFingerprintAuthenticationService,
            MyOwnNamespaceAssembly""/>

  </services>

</CompositeUI>
```

Note that in both cases, we have to explicitly remove the existing implementation before adding the new one. If we add a new one without removing the old one, perhaps hoping to overwrite it, CAB considers that a mistake and throws an exception.

I like to add services via config files during development for flexibility. But in most production code, that flexibility can become a potential source of errors, a hindrance rather than a help, when users or administrators mess with a file that they shouldn't have. So unless you need the flexibility at runtime, for example, to produce different configurations for different customers, then I suggest that you hard-wire them into code for production. If it is not necessary to make them flexible, then it IS necessary to make then NOT flexible.

5. Now we'll use the principles that we learned in modifying the prefabricated CAB services to write our own services. Most applications will contain at least a few. It's a very convenient mechanism for distributing programmatic logic in a loosely coupled way.

To demonstrate this service mechanism, I wrote my own CAB service. I first needed to define the interface that this service would provide. For simplicity, I provided just one method. Because it will be used by both the implementer of the service and the consumer of the service, I put it in a separate assembly named *TimeService.Interface.dll* to avoid any type of code dependency. Thus:

```
// Interface to be exposed by my new service

public interface ITime
{
    DateTime GetTime();
}
```

Now I needed to implement the service. It is common to put your implementation in one DLL and your interface definition in another. That way, you have to deliver the latter only to other developers, so they can't inadvertently set a reference to the implementation rather than the interface. In the case, I called that DLL *TimeService.dll*. That implementation looks like this:

```
public class TimeService : ITime
{
    public DateTime GetTime()
    {
        return DateTime.Now ;
    }
}
```

In addition to the techniques shown on the preceding page for adding services, you can also add a service by decorating the class definition with the attribute **Service**, which tells CAB to treat it in that manner. For this technique to work, the module in which it resides must be loaded by the CAB module loader service, which contains the code that recognizes the attribute. You can optionally specify the *AddOnDemand* attribute (not shown), which tells CAB not to instantiate it until someone asks for it. It looks like this:

```
[Service(typeof(ITime))]
public class TimeService : ITime
{
    <etc. >
```

6. In packaging my new service's declaration and implementation, I have several choices. I could put them into the SCSF–generated *Infrastructure.Interface* DLL (for the interface definitions) and *Infrastructure.Library* DLL (for service implementations). Because I have to roll these out anyway, I might as well piggyback my services along with those of SCSF. This would be a good choice if I had only a few services, and they were so central to my application that I always wanted them and never had to snap them out or snap other ones in.

Alternatively, I could generate a separate module to hold these services. The SCSF provides the choice of Foundational Module and Business Module. The primary difference is that the latter automatically creates and adds a new *WorkItem* to the chain, while the former does not. (Think "business = work" and you'll find it easy to remember.) Either of these is fine for a service. Since we want interfaces in a separate DLL, I'd probably select the Create an Interface Library for This Module check box,

This causes SCSF to generate a separate module for interfaces, as shown in Figure 2-5.

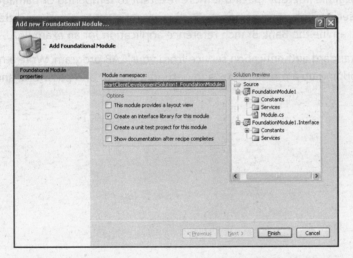

FIGURE 2-5 Separate implemenation and interface modules for a service.

Chapter 2 Lab Exercises
The Shell and Services

1. Generate a project using SCSF.

2. Add a class library to your solution. Make your application load it by making entries in your profile.xml file. Place any sort of demonstration method on your class that you like.

3. Turn your class library into a CAB service. Experiment with the different ways of adding it to your work item and think about the advantages and disadvantages of each one.

4. Add a module to your class. In your ModuleInit method, obtain a reference to the service. Try all the different methods you can think of—for example, the injection constructor or a regular service dependency.

5. Give some thought to how you would place your module load list into a Web service instead of a local file, thereby making it more resistant to tampering or damage by a thoughtless user, though perhaps not as much more resistant to a bad guy attack than you might think. Examine the Bank Branch reference application for an example.

6. Remove the standard authentication service from your CAB program. Write a new one, maybe just popping up a message box and getting a user ID and password, and plug in yours instead.

Chapter 3
WorkItems and Controllers

A.   Problem Background

1. The term *loosely*, as in *loosely coupled applications*, doesn't mean *neglected* or *badly*. A loose coupling mechanism needs as much careful thought and design as a tight one, perhaps even more. For example, how does an object that wants to display a view obtain a reference to the workspace in which the view should be shown? How does a subscriber to a CAB event discover which events are available and enter a new subscription? And how does code that wants to modify the contents of the main menu and toolbar find the locations at which these modifications are allowed and obtain a reference to the objects that perform these modifications? These are the sorts of problems that our loose coupling mechanism needs to solve.

This problem is similar to the dead-drop communication strategies so often used in espionage novels. The two parties don't want to know each other directly so as to allow easy replacement of either party with a new one that provides the same services. So, instead of meeting directly, the sending party writes a message and, at a specified time, puts it in a location agreed upon in some out-of-band specification. The receiving party checks that location when its circumstances allow and retrieves the message.

That's the sort of thing that we need to do in our loosely coupled applications. The code that creates a workspace needs to place it into a standardized location so that the code that uses the workspace can find it. The creator code doesn't know or care who will consume it, and the consumer code doesn't know or care who created it.

Loosely coupled applications need a standardized mechanism that allows creators of resources to place those resources into locations where their consumers can find them and where consumers can go to find resources placed by their creators, with neither party needing to have direct knowledge of the other.

B.   Solution Architecture

1.   A work item, represented by the class *WorkItem*, is the basic unit of software scoping in CAB. It contains well-known collections of objects such as services, workspaces, and Smart Parts, so that loosely coupled objects know where to find the resources that they need to do their work. It represents the primary mechanism whereby the loosely coupled parts of a CAB application are stitched together. In the preceding chapter, you saw the loosely coupled way in which a *WorkItem* contains services, making them available to other parts of a program.

The philosophy and purpose of the *WorkItem* have evolved greatly since its origin. The original idea was that a *WorkItem* object represented a particular use case—that is, one job of work, say a particular credit card operation. You would derive a new class from the base *WorkItem* and place business logic methods on it, as is seen in much of the early CAB samples and documentation. But it soon became apparent that (a) this granularity was too fine-grained because *WorkItems* were multiplying unnecessarily, (b) the resources held by a single *WorkItem* were often of interest to many instances of the same sort of use cases, and (c) the derived *WorkItem* class became too complex to effectively develop and test.

The current SCSF philosophy uses the *WorkItem* only as a container of other objects. You can think of a *WorkItem* as a scoping container in the same way as a Win32 process or a CAB *AppDomain* is a scoping container. The boundaries between *WorkItems* are easier to traverse than those others, but if you think of a *WorkItem* as a miniature subprocess or subapplication, you'll have the right mental model.

The business logic of a *WorkItem* now lives in a separate class called a *controller*, which the *WorkItem* contains. The relationship between the *WorkItem* class and its business logic has changed from "is-a" to "has-a." I discuss controllers in the last section of this chapter.

WorkItems are arranged in a tree-like chain, as shown on the facing page. The root work item, the base of the chain, is created by the shell application as part of its start-up process, as you saw in the preceding chapter. When modules are loaded into the process, they create work items and add them as children to the root work item's chain. The code for this is shown in section D of this lesson. Sometimes a module adds more than one work item, as does Module B in the diagram.

The *WorkItem* class contains properties that allow you to traverse this tree. The property *Parent* returns the parent *WorkItem*, and the property *RootWorkItem* takes you directly to the top of the chain. The child *WorkItems* of a *WorkItem* (in the diagram, 1 and 2 are children of the root, and 3 and 4 are children of 2) are located in the collection named *WorkItems*.

For the lack of a better word, and to use a highly overloaded one, the *WorkItem* to which an object belongs represents that object's context, the place in which it lives, the sea in which it swims.

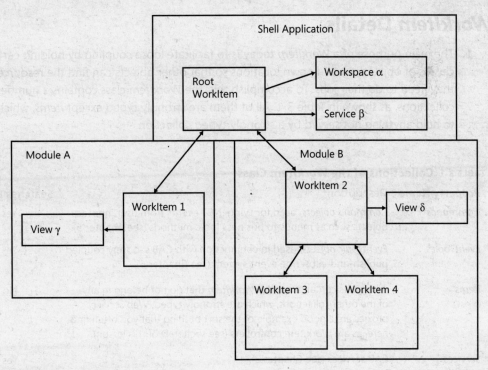

FIGURE 3-1 WorkItem tree.

C.   *WorkItem* Details

1. The main purpose of a *WorkItem* today is to facilitate loose coupling by holding certain classes of objects in well-known locations so that other objects can find the resources that they need to do their jobs. To accomplish this, the *WorkItem* class contains a number of collections, as shown in Table 3-1. All of them are strongly typed except *Items*, which exists to hold anything not served by a strongly typed collection.

TABLE 3-1 Collections of the *WorkItem* Class

Property Name	Description	Searches Parent
Commands	*Command objects*, used for tying .NET events from user interface objects such as menus to business logic methods (see Chapter 5).	Yes
EventTopics	*EventTopic objects*, used in conjunction with CAB's loosely coupled publish-and-subscribe event system (see Chapter 6).	Yes
Items	All objects contained in this *WorkItem* that do not belong in any of the other collections, which are strongly typed. Web service proxies are a good example of the sort of thing that you often find here, as are WorkItem controllers (see section D of this lesson).	No
Services	CAB services (see Chapter 2).	Yes
Smart Parts	Smart Parts, which are objects with the *SmartPart* attribute applied to them, also known as views (see Chapter 4).	No
State	*State* collection associated with this *WorkItem*.	No
UIExtensionSites	*UIExtensionSite* objects, used for negotiating the content of shared user interface items such as menus and toolbars (see Chapter 5).	Yes
WorkItems	Child *WorkItems* that belong to this *WorkItem*.	No
Workspaces	Workspace objects, which are frames used for visual display of views (see Chapter 4).	Yes

2. Objects are created and placed into the collections of a *WorkItem* in various ways. Sometimes this happens because of declarative attributes that are placed onto a class or a method. For example, the following code appears in the *ShellForm* class generated by the SCSF Wizard:

```
[EventSubscription(EventTopicNames.StatusUpdate,
    ThreadOption.UserInterface)]

public void StatusUpdateHandler(object sender, EventArgs<string> e)
{ … }
```

The *EventSubscription* attribute, when processed by the CAB framework, tells the framework to find the *EventTopic* object whose name is *StatusUpdate* in the *WorkItem* causing the load, creating it if necessary, and to add this particular subscription to it. That *EventTopic* is created and added to the *WorkItem* collection the first time this attribute is encountered by the CAB framework processing.

Other types of objects are added to *WorkItems* transparently when the CAB object builder creates them. In the *ShellForm* created by the SCSF, the object builder notes when it is creating the shell form that some of the child controls on the main form are *Workspaces* and not just plain windows. After creating them, the object builder places them into the root work item's *Workspaces* collection.

In other cases, objects are added to a *WorkItem* collection in response to explicit method calls. For example, in Chapter 2, we saw that services are often added to *WorkItems* in this manner. In section D of this chapter, you see how child *WorkItems* are added this way, and in Chapter 4, you see how Smart Parts are added in this way. Most often it's done by calling the AddNew< > method on the specific collection to which we want to add the new object. This tells the *WorkItem* to create the object, perform CAB processing on it, such as dependency injection, and return a reference to the newly created object. Alternatively, sometimes you choose a two-step creation process by creating the object separately with the *new()* operator and then calling Add<> to place it into the *WorkItem*.

Occasionally, an object is added to a *WorkItem* collection through a method call that's not quite as obvious as AddNew<>. For example, the call to *UIExtensionSites.RegisterSite*, which registers a particular user interface item, such as a menu for extension by other parts of the applications, also creates a new *UIExtensionSite* object in the *WorkItem*. (This technique is discussed in Chapter 5.)

3. To fetch an item from a *WorkItem* collection to use it, you usually use the Get method. When using the *Services* collection, as you saw in the preceding lesson, you specify the interface for which you want the implementing service, thus:

```
private void AuthenticateUser()
{
    // Fetch the authentication service from the root work item,
    // and call the method to make it authenticate the user.
    IAuthenticationService auth =
        rootWorkItem.Services.Get<IAuthenticationService>(true);
    auth.Authenticate();
}
```

The objects in the other collections are designated by string names. You fetch the desired item by indexing the collection with the string name, thus:

```
// User clicked this button. Look at event topic and
// fire it programmatically.
private void toolStripButton3_Click(object sender, EventArgs e)
{
    EventTopic et =
        _rootWorkItem.EventTopics[EventTopicNames.StatusUpdate] ;
    if (et != null)
    {
        et.Fire(this, new EventArgs<string>("Fired programmatically"),
            _rootWorkItem, PublicationScope.Global );
    }
}
```

As we've seen, *WorkItems* in a CAB application are arranged in a hierarchical chain. Sometimes a requested object in a collection is present at a higher level than the code making the request. For example, consider the *EventTopics* collection. The *StatusUpdate* event, owned by the shell form and used to update the status bar, can reasonably be fired from any object anywhere, which is why it's implemented with a loosely coupled CAB event and some tightly coupled technique such as a global variable. An object that has access to a child *WorkItem* two levels below the root could very reasonably want to fire this event, for which it needs access to the *EventTopic*.

Rather than force the requesting object to iterate up the *WorkItem* chain until it reaches the top, some of the collections in the *WorkItem* do this automatically. If an object asks *WorkItem 4* for the *EventTopic* named *StatusUpdate* (which lives in the root), it isn't found there, but the request is automatically sent upward to its parent (*WorkItem 2*). It isn't found there either, so the request is automatically sent up to *WorkItem 2*'s parent, which is the root *WorkItem*. In this collection, it is found and returned to the requesting object. This process happens automatically, and the requesting object isn't aware of it.

For other collections, searching up the *WorkItem* chain would not make sense. Consider the collection *WorkItems*, which really means "child *WorkItems*." If *WorkItem 2* in the preceding example wants to iterate through its children, it wants to find only *WorkItems 3* and *4*. It certainly doesn't want to find its siblings (*WorkItem 1*) or its ancestors (the root *WorkItem*). Therefore, some collections in a *WorkItem* search the parent chain and others do not. The "Searches Parent" column in Table 3-1 indicates which collections search upward through the chain and which don't.

4. One of the things that people like to do with any sort of container is to throw it away with all its contents in one neat operation (a kitchen trash bag springs to mind here). The *WorkItem* class provides this capability by implementing the *IDisposable* interface. If you call *Dispose* on a *WorkItem*, it iterates through all its collections and calls *Dispose* on every item that supports that interface. In the *CabApplication* class startup code, for example, the **Run** method calls *Dispose* on the root *WorkItem* to clean it up as the application is terminating. Thus:

```
// CabApplication class startup method

public void Run()
{
    < other initialization stuff … >

    // This method starts the message loop that runs the Windows Forms
    // application. It doesn't return until the app starts shutting
    // down.

    Start();

    // When this method returns, the app is shutting down. Dispose of
    // the root WorkItem, which disposes of all of its contents,
    // including any child WorkItems (and their contents, etc.).

    rootWorkItem.Dispose();

    // Dispose of the Visualizer too, while we're at it, because it
    // doesn't belong to any WorkItem.
    if (visualizer != null)
        visualizer.Dispose();
}
```

D. Controllers

1. To fulfill its function as a container, a *WorkItem* needs a place to hold its business logic. In the early days of CAB, this was often done by deriving a new class from *WorkItem*, such as *BankTellerWorkItem*, and placing the business logic on that class. Experience with the first CAB applications showed that this pattern allowed the *WorkItem* class to get too complex. Everything that a logical subapplication had to do or to hold was concentrated in just one class. It became the "God object," too large and complex to test or inspect, or even develop properly.

   By the time SCSF came out, the design philosophy had shifted to recommend separating the holding functionality from the doing functionality. Since the *WorkItem* already contained collections for holding objects, the idea emerged that it should store its business logic in a separate class, which came to be called the controller of the *WorkItem*. The holding and the doing thus resided on different classes in the same way that the Win32 operating system provides a process for holding things and a thread for doing things. Placing the business logic in its own separate class made things much easier to understand, especially for inspecting and testing.

   This relationship between the controller (the business logic object, remember) and the *WorkItem* (the resource container object) was codified in the class **ControlledWorkItem.** SCSF creates this class in the *Infrastructure.Interface* project. Shown on the facing page, it represents a *WorkItem* that contains a controller, and it's actually pretty simple.

   ControlledWorkItem derives from the *WorkItem*, so it contains all that considerable functionality. It is sealed, which means that it cannot be used as the basis for any further derivation. This is strong hint to developers that the buck really ought to stop here.

   You can see that the *ControlledWorkItem* class contains a controller object, which it creates when its **OnBuiltUp** method is called. Using the **AddNew< >** method of the *WorkItem* causes the new controller to receive CAB processing, such as dependency injection. The class of this controller object is passed as a generic parameter, called *TController* in the example. This controller class varies from one instance of a *ControlledWorkItem* to another, but the *ControlledWorkItem* that contains it is always identical to all other *ControlledWorkItems*. The controller is stored internally in a member variable and accessed through a read-only accessor property in the usual way.

```
public sealed class ControlledWorkItem <TController> : WorkItem
{
    private TController _controller;

    // Accessor property for the WorkItem's controller object

    public TController Controller
    {
        get { return _controller; }
    }

    // When the WorkItem object is finished being constructed
    // internally, then create the controller object. Save a reference
    // to it in a member variable, and also place it into the Items
    // collection of the WorkItem.

    public override void OnBuiltUp(string id)
    {
        base.OnBuiltUp(id);

        _controller = Items.AddNew<TController>();
    }
}
```

2. The creation of a *ControlledWorkItem* and the most common means of connecting it to the *WorkItem* chain are shown on the facing page. When the CAB module loader service loads a module, it uses the reflection API to examine the module's assembly for classes that derive from **ModuleInit.** When it finds such a class, the loader instantiates it. The SCSF Wizard generates modules with an injection constructor asking for injection of the root work item, which it stores in a private variable. Finally, the module loader service calls the *ModuleInit* object's Load method.

In the Load method we instantiate the *ControlledWorkItem* object by calling the method AddNew<> on the root work item, passing the class *ControlledWorkItem* as a generic parameter. This tells the root work item to create a new object of the specified class and place it into its *WorkItems* collection. Creating the *ControlledWorkItem* causes it to create its internal controller, as we saw on the previous page.

After we create the new *ControlledWorkItem* with its controller, we need to initialize it. The *WorkItem* is already built up internally, but we haven't done anything yet with its business logic, which resides on the controller. We therefore call the controller's Run method, thereby telling it to jump up and start doing its thing, whatever that might be. Let's now turn our attention to the controller class itself.

Note The name of the variable in the injection constructor is *rootWorkItem*, but this is sometimes a misnomer. The *WorkItem* that is passed is the *WorkItem* that caused the module to be loaded, passed as the first parameter to the module loader service's Load method. This happens most frequently in the normal application startup method *CabApplication.LoadModules*, where it is indeed the root *WorkItem*. However, it is possible, though somewhat unusual, for a child *WorkItem* to fetch the module loader service and use it to load modules (the *WorkItems* of which will be grandchildren of the root *WorkItem*). In this case, the child *WorkItem* will probably pass itself, not the root *WorkItem*, as the parameter to the Load method, in which case the injected *WorkItem* will be the child *WorkItem*, not the root. This parameter would be more accurately named *parentWorkItem*, even though the parent is almost always the root.

```
// The Module Loader service looks in the modules that it loads for
// classes that derive from ModuleInit. When it finds one, it
// instantiates it and calls Load.

public class Module : ModuleInit
{
    private WorkItem _rootWorkItem;

    // This injection constructor causes the root work item to be
    // injected into the new ModuleInit object. This allows the
    // WorkItem that this module will create to be connected to the
    // application's WorkItem chain.

    [InjectionConstructor]
    public Module([ServiceDependency] WorkItem rootWorkItem)
    {
        _rootWorkItem = rootWorkItem;
    }

    public override void Load()
    {
        base.Load();

            // Create a new ControlledWorkItem, placing it into the
            // RootWorkItem's collection of WorkItems. The process of
            // creating the ControlledWorkItem also creates an object
            // of class MyOwnModuleController, and attaches it to the
            // ControlledWorkItem.

        ControlledWorkItem<MyOwnController> workItem =
                rootWorkItem.WorkItems.AddNew<
                ControlledWorkItem<MyOwnController>>();

            // Call the Run method on the new WorkItem's Controller
            // so as to perform its initialization

        workItem.Controller.Run();
    }
}
```

3. The controller for a *WorkItem* derives from the SCSF base class **WorkItemController,** shown on the facing page. It is generated by the SCSF and lives in the *Infastructure. Interface* project. It obtains the *WorkItem* to which it is connected by means of dependency injection.

The *IWorkItemController* interface contains the single method Run(), used at startup time, as we saw on the previous page. Apart from that, a controller can contain any code. You will note that the *TController* type in the definition of the *ControlledWorkItem* class does not specify that *WorkItemController* be the base class or that the controller implement the *IWorkItemController*. To do that, we would specify a derivation constraint (see Appendix A) written as follows:

```
public sealed class ControlledWorkItem <TController> : WorkItem
  where TController: WorkItemController
```

The base class goes to the trouble of exposing the *ActionCatalogService* (see Chapter 7) as an accessor property. Although this service is sometimes useful, it's not clear to me that this service is important enough to expose it in this special case and while not exposing other services in a similar manner.

```csharp
// Base class for controller user by ControlledWorkItem class

public abstract class WorkItemController : IWorkItemController
{
    private WorkItem _workItem;

    // This dependency injection obtains for the controller a reference
    // to the WorkItem that contains it.

    [ServiceDependency]
    public WorkItem WorkItem
    {
        get { return _workItem; }
        set { _workItem = value; }
    }

    public IActionCatalogService ActionCatalogService
    {
        get { return _workItem.Services.Get<IActionCatalogService>(); }
    }

    public virtual void Run()
    {
    }
}
```

4. A sample controller class is shown on the facing page. The SCSF Wizard generates it with the class name *ModuleController*, but I find that a misnomer. The controller isn't tied to the module; it's tied to a *WorkItem*. The module is merely a convenient administrative container to facilitate being loaded by the module loader service. I find that changing the name to something that better describes what the controller's purpose is, such as *PharmacyController*, is more conducive to understanding as development proceeds.

The controller's **Run** method generally calls utility functions to do its internal setup, as shown on the facing page. This often entails adding services to the *WorkItem* (Chapter 2), creating and adding views and presenters to workspaces (Chapter 4), or modifying the shared user interface (Chapter 5). You certainly don't have to do it this way; you are free to modify it as needed, but simple cases might as well start out with it and see how far they get.

The controller class often contains methods that are command handlers for a shared user interface (Chapter 5) or event subscriptions (Chapter 6). It may also contain Action methods or conditions (Chapter 7) . Any code that is scoped at a *WorkItem* level, as opposed to an individual view or presenter, generally belongs on it.

```csharp
public class PharmacyController : WorkItemController
{
    public override void Run()
    {

        // Load any additional services that the item might need.

        AddServices();

        // Extend the shell application's menu, as explained in
        // Chapter 5.

        ExtendMenu();

        // Extend the shell application's toolbar and status bar,
        // as explained in Chapter 5.

        ExtendToolStrip();

        // Create and wire up the Smart Parts, Views, and Presenters
        // used for displaying the user interface, as described in
        // Chapter 4.

        AddViews();
    }
```

Chapter 3 Lab Exercises
WorkItems and Controllers

1. Generate a new SCSF project, or use the one you started with in the preceding project.

2. In your module, add a new *ControlledWorkItem* and controller class.

3. Using the sample visualizer program, examine the contents of your *WorkItem*.

4. In your controller, create a new child *WorkItem* and add it to the original one. Experiment to see which collections search the collections of their parent *WorkItem* and which do not.

Chapter 4
Workspaces and SmartParts

A.   Problem Background

1. The display of visual information is a constant problem in a loosely coupled application, where components are not developed together and in general don't know much about each other. A subordinate module needs to display its visual information, and the shell form needs to provide an area for this.

   Consider the Rolling Thunder Hospital sample application that we have been working with throughout this book, shown schematically in Figure 4-1. Not only do the developers of, for example, the X-ray module have intimate knowledge of the data and service silo(s) in which the patients' X-rays reside, but they also are the primary experts on the ways in which doctors like to view X-rays, which the developers of the outer shell probably are not. The same holds for the pharmacy and other modules. The developers of the subordinate modules are closer to the users and therefore know the right way to display prescriptions and funeral choices and so on. To place the shell developers in charge of displaying this information to the user would not only violate loose coupling, but it would probably lead to a bad user interface.

   The shell development team, for its part, is in charge of the overall look and feel of the application. The shell wants to have the minimum possible involvement in the internal logic of the modules, but it does need to provide them with a place to do their displays. It is the shell team that says, "OK, the patient selection controls are going to be on the left, because selecting a patient is the first operation that a user has to do. Then display of the individual data items will be on the right, and to accommodate different views in the available real estate and to customize the set of items that are shown to each user, we'll show each module's display in a separate tab."

   We need to have a way in which the shell team can make broad layout decisions such as these without affecting the implementation of the subordinate module's display. And we need a way in which the subordinate modules can display their visual information to the user in a manner not tightly coupled to the shell so that the user interface of each particular part can evolve independently.

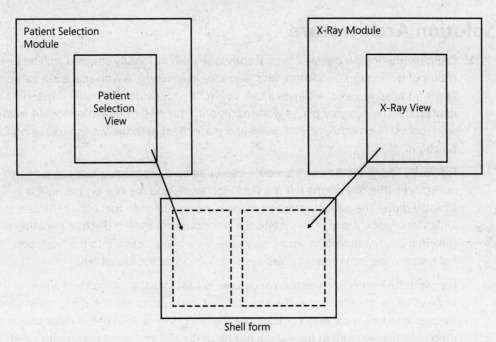

FIGURE 4-1 Schematic diagram of modules displaying Smart Parts in workspaces.

B.   Solution Architecture

1. CAB provides for the display of visual components in its loosely coupled architecture by means of *workspaces* and *SmartParts* (also known as *views*). A workspace is a frame, a container for hosting views. A view is a user control that provides the visual display of a work item's data. The designer places workspaces onto the shell form and optionally inside other workspaces. The modules create views and place them onto the workspaces using CAB functions.

   Figure 4-2 shows the Rolling Thunder Hospital sample program, which contains two workspaces. The one on the left is a *DeckWorkspace*, and the one on the right is a *TabWorkspace*. The patient selection module creates the view that allows the user to select an available patient and places it into the workspace on the left. Each of the other modules (pharmacy, X-ray, mortician, etc.) creates one view and places it into the workspace on the right, where the *TabWorkspace* displays each view in an individual tab.

   Even with CAB's mechanisms for workspaces to host views, a CAB project almost always requires some amount of out-of-band communication between the design teams to make the user interface work well. For example, the designers of each of the views shown in the preceding figures need to have some notion of the size and aspect ratio with which the view can expect to be displayed. They will lay out their child controls in a completely different manner if the hosting workspace is wide and narrow (landscape orientation) than they would if it were tall and deep (portrait orientation). And while the *SmartPartInfo* (see section E of this chapter) allows for some amount of negotiation at display time, it is essentially impossible to design a view that dynamically configures itself so as to look good in both a portrait and also a landscape orientation.

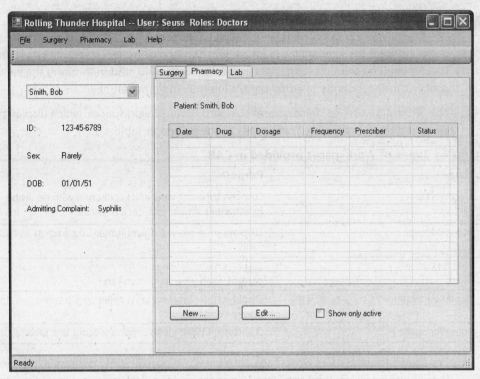

FIGURE 4-2 Rolling Thunder Hospital sample program showing workspaces.

C.   Workspaces

1.   A *workspace* is the container, the frame, in which one or more views is shown. It provides a standardized mechanism for displaying, hiding, activating, and monitoring the views that it contains, independently of either party's internal implementation.

The CAB framework contains several standard types of workspaces, which display their views in different manners. These types are described in Table 4-1.

TABLE 4-1 Types of Workspaces provided in CAB

Name	Purpose
DeckWorkspace	Displays topmost view in its collection with no visible frame, in a manner similar to a deck of cards.
MdiWorkspace	Displays multiple views simultaneously, each in an MDI child form.
TabWorkspace	Displays multiple views in the manner of a standard tab control, each view in its own tab.
WindowWorkspace	Displays topmost view in its collection, framed in a floating popup window.
ZoneWorkspace	Displays multiple views simultaneously, in a tiled layout.

Examples of each type of workspace are shown in Figure 4-3. My customers tell me that they find themselves using the *DeckWorkspace* and the *TabWorkspace* most frequently.

You can also develop your own workspaces without much trouble, as described later in this chapter.

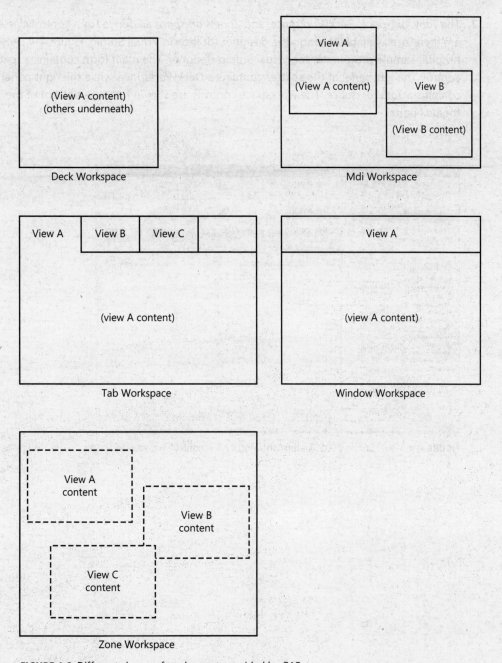

FIGURE 4-3 Different classes of workspaces provided by CAB.

2. The *TabWorkspace*, *DeckWorkspace*, and *ZoneWorkspace* all derive from control classes and are therefore available through the designer toolbox in Visual Studio. Figure 4-4 shows the hospital sample program in the Visual Studio designer. The main form contains a splitter control. The left panel of the splitter contains a *DeckWorkspace*, while the right panel contains a *TabWorkspace*. The workspace controls are shown near the bottom of the toolbox panel.

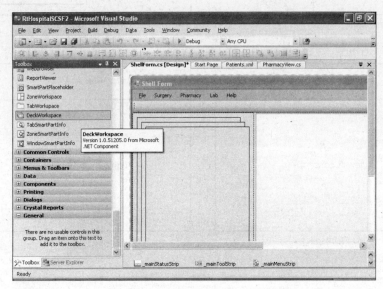

FIGURE 4-4 Visual Studio 2005 Toolbox showing CAB's control-based workspaces.

3. The *MdiWorkspace* and the *WindowWorkspace* classes don't appear in the toolbox because they don't derive from controls and hence don't contain the code that they need to be hosted there. Instead, you must write code to explicitly to create them. Here is an example of creating an *MdiWorkspace* and adding it to a work item's collection of workspaces:

```
private MdiWorkspace mainWorkspace ;
private WorkItem workItem ;

[InjectionConstructor]
public BankShellForm(WorkItem workItem) : this()
{
    this.workItem = workItem;

    // Create new MdiWorkspace

    mainWorkspace = new MdiWorkspace(this);

    // Place it in the work item's collection

    this.workItem.Workspaces.Add(mainWorkspace, "mainWorkspace");
}
```

4. All workspaces support the *IWorkspace* interface. This interface allows a client to add, remove, show, and hide views in the workspace. Its members are shown in Table 4-2.

TABLE 4-2 Members of the *IWorkspace* Interface

PROPERTIES:

Name	Function
ActiveSmartPart	The view that is currently active in the workspace. Each type of workspace has its own implementation of visual properties that indicate which view is active—for example, the *TabWorkspace* showing the tab containing the active view on top of the other tabs.
SmartParts	Read-only collection of all views currently contained in the workspace.

METHODS:

Name	Function
Activate	Makes the specified view the workspace's active view. The view must already exist in workspace's collection.
ApplySmartPartInfo	Applies the specified *SmartPartInfo* (see section E of this chapter) structure to the specified view.
Close	Removes the specified view from the workspace's collection. Disposing the view is the responsibility of the caller.
Hide	Makes the specified view invisible in the workspace but leaves it in the workspace's collection.
Show	Adds the specified view to the workspace's collection if it isn't already there. Then internally calls the workspace's *Activate* method, unless explicitly directed otherwise by a provided *SmartPartInfo* object.

EVENTS:

Name	Function
SmartPartActivated	Fired when a SmartPart is activated.
SmartPartClosing	Fired when a SmartPart is closing.

For graybeards like me, this interface is roughly analogous to the good old *IOleInPlaceSite* interface back in Object Linking and Embedding (OLE).

5. What a client most often wants to do with a workspace is create views and show them in that workspace. To do that, the client needs to figure out which workspace he wants to show the view in and obtain a reference to that workspace. This can be done in two basic ways.

The pattern used by most of the samples that come with CAB, and most of the industrial applications I've seen and heard about, is the "well-known name" pattern. The shell designer writes, in some sort of out-of-band spec, that the shell will contain workspaces with certain string names. She then ensures that the root work item contains workspaces with these names, either by placing them on the shell form at design time or by explicitly creating them through code. The code that creates and shows the view uses this name to fetch the *IWorkspace* interface from the work item's collection and then uses this interface to add the view. The June 2007 release of the SCSF contains the method *ShowViewInWorkspace* (not shown) that performs both of these operations in a single call. Thus:

```
private AddViews()
{
   MyView mv ;

   //Retrieve workspace by means of its well known name

    IWorkspace mainWorkspace =
        this.workItem.Workspaces[WorkspacesConstants.MAIN_WORKSPACE];

   // Tell the WorkItem to create the view that we want to show in the
   // workspace, and place it into the WorkItem's SmartParts
   // collection

   mv = this.WorkItem.SmartParts.AddNew <MyView> () ;

   // Show the view in the main workspace

   mainWorkspace.Show(mv);
}
```

This well-known name pattern is suitable for the vast majority of cases. However, occasionally the situation is so dynamic or complicated that static out-of-band communication doesn't solve the problem. In this case, you have to write a CAB service for figuring out which workspace ought to be used by whom and when and for what purpose. The code that wants to display the view will fetch this service using the *Services* collection of the *WorkItem* (shown in Lessons 2 and 3) and call whatever method you have provided on it, passing whatever parameters you have deemed necessary. The service's code will then figure out which workspace the client should use and returns a reference to it or perhaps the string name for allowing access through the *Workspaces* collection of the *WorkItem*.

6. You can implement your own workspace classes with surprisingly little effort. CAB provides a class called **WorkspaceComposer** that provides their basic functionality. You can derive a workspace directly from this, but a more common case is to graft workspace behavior (support for the IWorkspace interface, which provides the ability to host views) onto a control that you already have and like. In this case you can't change the base class. Instead, your derived class implements the **IComposableWorkspace** interface and delegates internally to the *WorkspaceComposer* object (hereafter called the *composer*). This is how the *TabWorkspace* and others provided in the CAB libraries are implemented internally.

To demonstrate writing your own workspace, I wrote a sample that takes the Windows Forms *FlowLayoutPanel* control and turned it into a CAB workspace. Some of the code, developed fully in the sample project *FlowLayoutWorkspace*, is shown on the facing page.

First, you derive your own control class from the one you want to emulate, in this case, *FlowLayoutPanel*. You make it implement the *IComposableWorkspace* interface, which derives from *IWorkspace*. In the constructor, you create the composer object. Then you add *WorkItem* injection, passing the injected *WorkItem* to the composer, because the composer needs access to the *WorkItem* chain. So far, nothing difficult.

Next, let's implement the methods of the *IWorkspace* interface. The composer contains prefabricated functionality for doing exactly this. Using it not only saves us time and trouble, but also means that our workspace's behavior will be consistent with that of Microsoft-provided workspaces because they use this class internally. We delegate all these methods to the composer. The facing page shows only the **Activate** method.

We still need to implement the remaining methods of the *IComposableWorkspace* interface which deal with the connection from the composer to the control from which we derive. The composer doesn't, can't, know what it means for our workspace to host a view—for example, which method to call on the base class to add or remove a view. When the client calls *IWorkspace.Show*, telling it to add a new view, the composer calls our method IComposableWorkspace.OnShow. When the client calls *IWorkspace.Close*, telling it to close an existing view, the composer calls our method IComposableWorkspace.OnClose. It's up to our code to take that new view and do whatever it is that our base control needs done in order to show it (in the former case) or get rid of it (in the latter case). In the sample code, we know that our base class handles these operations with its own internal collection called Controls. The only code that our implementation needs to provide is the bridge between the composer and this collection, as you can see on the facing page. Other methods, such as OnHide and OnActivate, are called by the composer when it needs to, the former meaning to make the view invisible and the latter to make it visible if necessary and to make it the active view. Our code (not shown here but available in the sample project) knows what that means in terms of our workspace's base class and performs these operations.

The composer does much of the scutwork of the *IComposableWorkspace* interface, such as checking for null parameters. For another example, when the client calls *IWorkspace.Show*, passing a view, the composer checks to see if that view already exists in its collection. It then calls *IComposableWorkspace.OnShow* if the view is new and needs to be added to our workspace, or *IComposableWorkspace.OnActivate* if the view already exists in our collection and simply needs to be brought to the foreground. It also maintains the *SmartParts* collection of the *IWorkspace* interface.

```
// Derive our workspace class from a non-workspace control, implement
// the IComposableWorkspace interface

public class FlowLayoutPanelWorkspace : FlowLayoutPanel,
    IComposableWorkspace <Control, SmartPartInfo>
{

    // This is the CAB-provided WorkspaceComposer object, which
    // implements most of the functionality of a workspace

    private WorkspaceComposer<Control, SmartPartInfo> composer;

    // In the class constructor, create the composer as well

    public FlowLayoutPanelWorkspace()
    {
        composer = new WorkspaceComposer<Control, SmartPartInfo>(this);
    }

    // Ask for the current WorkItem to be injected. Pass the WorkItem
    // to the composer.

    [ServiceDependency]
    public WorkItem MyWorkItem
    {
        set { composer.WorkItem = value; }
    }

    // For this and all the IWorkspace methods, simply delegate to the
    // composer, which exists for this purpose

    public void Activate(object smartPart)
    {
        composer.Activate(smartPart);
    }

    < ... other IWorkspace methods >

    // For this and all IComposableWorkspace methods, we have to write
    // code that mediates between the composer's collection of views
    // and our workspace's mechanism for hosting and displaying them.
    // In the case of a FlowWorkspacePanel, that means putting a new
    // view into the controls collection or taking it out again.

    public void OnShow(Control smartPart, SmartPartInfo smartPartInfo)
    {
        this.Controls.Add(smartPart);
    }

    public void OnClose(Control smartPart)
    {
        this.Controls.Remove(smartPart);
    }
}
```

D.   SmartParts (Views)

1.   A SmartPart or view is a visual component of a CAB application, displayed in a workspace, usually providing a user interface to some business logic contained in a work item. The terms *SmartPart* and *view* are synonymous and used interchangeably. I've settled on the latter in this book because it's shorter and doesn't try to be cute. A view is usually based on a Windows Forms user control.

You generally use a view in conjunction with a presenter, the Model-View-Presenter design pattern you've heard so much about. The arrangement of components is shown in Figure 4-5. This organization is conceptually similar to the classic three-tier architecture used in much of distributed programming, although in this case, it is being used to organize the layers of a client application.

The view is conceptually similar to the presentation layer of the three-tier stack. So as not to confuse it with the presenter, which represents the middle tier, I describe the view as the *display layer*. The view directly interacts with the user, displaying data and accepting input. It forwards almost all user input to the presenter for further cogitation. The only user input that a view handles within itself would be input related solely to the current display of the view, such as a zoom factor. Even a color preference would be forwarded to the presenter for storage in the model layer, if it were intended to persist from one user session to another.

The middle layer, conceptually similar to the business tier in the classic three-tier architecture, is called the *presenter*. It mediates between the view and the rest of the program, performing the client application's business logic. In the hospital sample program, the pharmacy presenter examines the user's credentials and populates the pharmacy view's *AvailableDrugs* properties with only those drugs that the user is authorized to prescribe.

The presenter does not have direct knowledge of the class to which the view belongs. Instead, it communicates with the view solely through an interface exposed for this purpose. Since user interface designs often change dramatically during development, and often in production-time configuration from one customer to another, we want to confine those changes to the view class. If the presenter had intimate knowledge of all the items in a view, and you then changed the display of string from a label to a text box, you would have to change not just the view code, but the presenter code as well. We don't usually bother isolating the presenter from the view in the same manner. The view usually knows the class of its presenter and makes calls directly on it. The presenter's public methods exist to serve the view, so they shouldn't be changing for external reasons.

You add a view, with its associated interface and presenter, to your SCSF project using the Add View (with Presenter) menu item. The files are generated as shown in Figure 4-5, as illustrated in the sample code project *SimplestViewPresenter*. We next discuss each of them in detail.

FIGURE 4-5 View added to SCSF project in Visual Studio 2005.

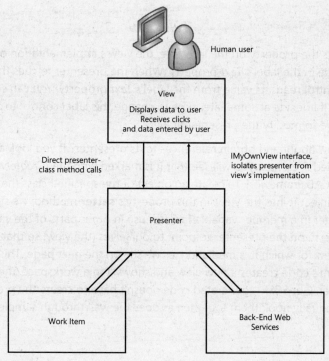

FIGURE 4-6

2. The code for a view is shown on the facing page. It derives from the base class *System.Windows.Forms.UserControl*, so it contains all that considerable functionality. It implements the *IMyOwnView* interface, defined in a separate file. This interface contains the data that the control wants to expose to its presenter. In this case, I've added a single string property called *Message*. The definition of the interface is as follows:

```
public interface IMyOwnView
{
    string Message { set; get;}
}
```

To implement the interface, I've added a label control to the *UserControl*, thus:

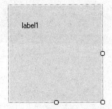

When the presenter sets the property on the interface, the view's implementation of that property stores the value in the label's *Text* property. When the presenter fetches the value of this property, the control reads its value from the label's *Text* property. You can see the isolation working here. If I decide at some later time to change this label control to a text box instead, I won't have to modify the presenter at all.

How does the view know about and obtain references to its presenter? If you look at the public property called *Presenter*, you will see that it is marked with the *[CreateNew]* attribute. This tells the CAB Framework to create an object of the specified class (here *MyOwnPresenter*) and inject it into the view. In this property's **setter** method, we store the value of the presenter in a member variable for later use in other parts of the view. We then set the *View* property on the presenter to point to ourselves (the view) so that the presenter knows the view for which it is mediating, as we see on the next page. This means that when our work item's code creates a new view and shows it in a workspace, the view creates its own presenter. Our SCSF—generated code doesn't have to create the presenter separately and then manually hook them together, as does the walkthrough sample application code.

```
[SmartPart]
public partial class MyOwnView : UserControl, IMyOwnView
{

    // This attribute tells the dependency injection system to
    // automatically create a new presenter and hand it to us. We
    // store a reference to it locally so we can talk to it later.
    // We also set the presenter's View property, which is how it
    // finds out which view it is working with.

    [CreateNew]
    public MyOwnViewPresenter Presenter
    {

        set
        {

            _presenter = value;
            _presenter.View = this;

        }
    }


    // Tell the presenter that the view is up and ready for business.

    protected override void OnLoad(EventArgs e)
    {
        _presenter.OnViewReady();
    }

    // Implement the Message property on the IMyView interface

    public string Message
    {
        get
        {
            return this.label1.Text;
        }
        set
        {
            this.label1.Text = value;
        }
    }
}
```

3.   The code for a presenter is shown on the facing page. It derives from the base class **Presenter< >** (not shown), which is generated by the SCSF and lives in the *Infrastructure. Interface* library project. This base class contains a property named *View*, which is of the interface type through which the presenter accesses its view (in this case, *IMyOwnView*). We specify that type through the generic parameter as shown. The value of this property is set by the view code, as shown on the previous page (see the property *Presenter*, in the *setter* method by which the presenter is injected). In the sample code, my presenter uses this interface to set the view's *Message* property.

The base class also contains a dependency injection property, asking for the *WorkItem* in which it resides. (I haven't shown it in this book, but you'll see it if you look at the base class's code.) So the connection of the presenter to the model is made for us automatically as well. Sometimes the presenter will use this connection to access other objects via the *WorkItem* (the lower-left corner of Figure 4-64). In other cases, the presenter will access its back-end data directly via web services to which only it has a reference (the lower-right corner of Figure 4-6), that is, which are not made available to other objects via the work item.

The presenter contains a method called OnCloseView( ), which is somewhat misnamed. It sounds like an event handler, but it isn't. It is the method that a work item's code calls on a presenter to tell the presenter to close its view. The base class's method searches the *WorkItem* chain for the workspace in which the view resides and calls its Close() method, thereby removing the view from the workspace. This means that you don't have to remember which workspace contains the view to be able to close it; you can instead simply ask the presenter to do so. This codes uses the WorkspaceLocatorService (not shown), generated by the SCSF, which resides in the *Infrastructure.Library* project.

```
public class MyOwnViewPresenter : Presenter<IMyOwnView>
{
    // This method will be called by the view when it's been loaded

    public override void OnViewReady()
    {
        base.OnViewReady();

        // Use our interface to the view to set the view's
        // Message property.

        this.View.Message = "Hello, world";
    }

    // Close the view

    public void OnCloseView()
    {
        base.CloseView();
    }
}
```

4. You probably noticed that the view class is marked with the attribute *[SmartPart]*. The function of this attribute is not immediately clear because the views marked with it superficially seem to behave in a manner similar to those that are not. For example, this attribute is not required for a control to be hosted in a workspace. Try it for yourself, comment it out, and you probably won't notice any immediate change in simple application behavior. But it does have an explicit and important purpose, even if that purpose is not immediately clear. This attribute does two things and these two things only, as demonstrated in the sample program *SmartPartAttributeMeaning*.

First, when a view marked with this attribute is created via the AddNew( ) method of the *WorkItem*, the view is stored in the *SmartParts* collection of that *WorkItem* (see Figure 4-7). A view that lacks this attribute does not go into the *SmartParts* collection, but rather into the *Items* collection, which is used for objects of types for which the *WorkItem* does not provide a strongly typed collection (see Figure 4-8). The view itself will display correctly, but any code that looks for the view in the *SmartParts* collection will not find it there. For example, the *WorkspaceLocatorService*, used in the **OnCloseView** method shown on the previous page, examines only the *SmartParts* collection. If the view is not marked with this attribute, then the service will not find it and *OnCloseView* will not work the way that you expect. This function of the attribute can be logically phrased as "Always store me in the *SmartParts* collection of the *WorkItem* rather than its *Items* collection."

The second meaning of the attribute occurs when a view is created not by an explicit call to a *WorkItem,* as in the previous case, but when a view is created declaratively by being placed as a child control on another control. Suppose a view of class A contains a child view of class B (see Figure 4-9), placed there using the designer. Suppose further that view B requires dependency injection—perhaps it needs to have its presenter automatically created as did view A, or perhaps it's not bothering with a presenter but needs its *WorkItem* injected so as to find an *EventTopic* for firing a loosely coupled CAB event (see Chapter 7). If view A is created through an explicit call to the *WorkItem*, as in the previous paragraph, view A always receives CAB processing and dependency injection, whether it is marked with the *SmartPart* attribute or not. However, view B, being created by as a child control of view A, does NOT receive CAB processing unless it is marked with the *SmartPart* attribute. The object builder, in constructing view A, examines its children and performs CAB processing on those that are marked with this attribute and omits it from those that are not. This function of the attribute can be logically phrased as "Perform CAB processing on me when I'm created as the child of another CAB object."

The SCSF recipe generates views that are marked with this attribute, and you usually leave it on. But once in a while I am asked, "What would happen if you took it off?" Now you know and can prove it to yourself by running the sample application in every permutation. You might as well leave it on.

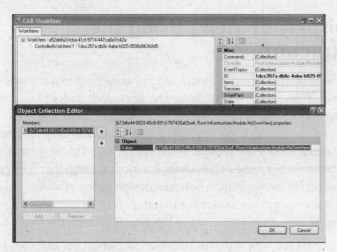

FIGURE 4-7 CAB Visualizer showing view in SmartParts collection because it is marked with the *[SmartPart]* attribute.

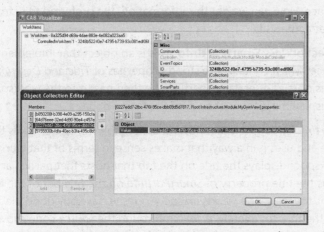

FIGURE 4-8 CAB Visualizer showing same view in Items collection because it is NOT marked with the *[SmartPart]* attribute.

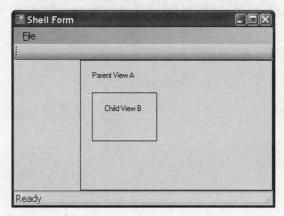

FIGURE 4-9 One CAB view containing another CAB view as a child control.

E.   Modifying the Display of Views Using the *SmartPartInfo* Structure

1.   The connection that we've seen between the view and the workspace that displays it is quite loose, very much an arm's length sort of thing. The workspace doesn't know anything about how the SmartPart would like to be displayed—for example, what its preferred size or caption might be. The view doesn't know anything about what its display options are, such as whether or not it can specify the caption or the display size. This very loose style of connection is often acceptable, since the details change seldom enough that they can be handled through out-of-band communication between design teams. However, situations arise in which we want the view and the workspace that displays it to exchange information about each other at display time, so as to customize the display to a greater degree.

Consider the *TabWorkspace*. Each view is shown in its own tab, which the user switches among while performing his work. Since the contents of the tab are provided by the view, it makes sense that the title should also be provided by the view, but the tab and its title are owned by the workspace. How can we get this information, the preferred title, from the view to the workspace that contains it?

The answer is with an object of class **SmartPartInfo.** This is an object that implements the *ISmartPartInfo* interface, which contains the string properties of *Title* and *Description*. We create an object of this class and set its properties. We then pass it to the workspace's *Show* method when we display the view, as shown on the facing page.

When we pass a *SmartPartInfo* structure while showing a view, the workspace takes the information in the structure and uses it in a way that makes sense in terms of that workspace. For example, the *TabWorkspace* displays the title on the tab that hosts that particular view. An *MdiWorkspace* displays the title property of *SmartPartInfo* on the title bar of the MDI child window that contains the view. The *DeckWorkspace* doesn't have any good place to show a title, so it ignores the title completely. A search through the source code shows that the *Description* property is not currently used by any workspace. It would appear to be provided for internal debugging or perhaps future compatibility.

```
private void AddViews()
{

    // Create the view

    MyOwnView mov = this.WorkItem.SmartParts.AddNew<MyOwnView>();

    // Create the SmartPartInfo for displaying view

    SmartPartInfo spi =
        new SmartPartInfo("MyOwnTitle", "MyOwnDescription");

    // Show the view in the workspace using the SmartPartInfo structure

    this.WorkItem.Workspaces[WorkspaceNames.LayoutWorkspace ].Show (

        mov,            // view to display
        spi);           // SmartPartInfo

}
```

2. Once you start thinking about customizing the runtime interaction of views and workspaces through the intercession of *SmartPartInfo* objects, you realize that this interaction may vary greatly from one class of workspace to another. For example, the WindowWorkspace is a free-floating popup window, so the view might have some preferences as to size, decorations such as a system menu or minimize box, etc. The size of a *TabWorkspace*, on the other hand, is probably set at design time, but the order of the tabs probably isn't. A view might reasonably want to add itself to the front of the tab collection or to the back or somewhere in the middle. How can we provide different types of information to the different classes of workspaces, and how can we discover at runtime what types of information a particular workspace is capable of handling?

We do this by deriving different classes of *SmartPartInfo* containing information customized to a particular type of workspace, for example, *TabSmartPartInfo*. In addition to the base class properties, this class adds properties named *ActivateTab* and *TabPosition*. This allows the view to have a certain type and amount of control over its display in a *TabWorkspace* that would not make sense in a different type of workspace. The former property tells the *TabWorkspace* whether or not to make the newly added view the active tab. The latter tells the *TabWorkspace* whether to put the new view onto the front of the tab collection (far left side, position 0) or the back of it (far right side). It doesn't provide a numeric index allowing you to stick it somewhere in the middle. Other classes of workspace also have their associated *SmartPartInfo* structures, such as *ZoneSmartPartInfo* and *WindowSmartPartInfo*, which apply to the *ZoneWorkspace* and *WindowWorkspace* classes, respectively. When you implement your own class of workspace, as described earlier in this chapter, you may decide to also implement your own derived class of *SmartPartInfo* that provides information used by that workspace.

How does a view know which type of workspace it is being hosted in and, therefore, which type of *SmartPartInfo* to provide? It can implement the **ISmartPartInfoProvider** interface, as shown on the facing page, which contains the single method **GetSmartPartInfo()**. When the view is shown in a workspace, the workspace queries it for this interface and calls the method if found, passing the type of *SmartPartInfo* that it knows how to use. A *TabWorkspace*, for example, passes the type object that describes *TabSmartPartInfo* [not a *TabSmartPartInfo* itself, but *typeof(TabSmartPartInfo)*]. It is up to the view to examine this type, figure out what it has to say to a workspace that wants that type of *SmartPartInfo*, and return the correct class. The return type is *ISmartPartInfo*, so if the view doesn't recognize the particular requested type or doesn't have anything meaningful to add to the base type, it can always return a plain old *SmartPartInfo*. In the sample on the facing page, the Rolling Thunder Hospital program is hosting the *MorticianView* in a *TabWorkspace*, so the request when it comes in will specify the *typeof(TabSmartPartInfo)*. However, the view doesn't care about the additional properties of the *TabSmartPartInfo*, since its position and activation in this application are being centrally managed. It therefore returns a simple *SmartPartInfo* containing its desired title.

```csharp
// This class implements the ISmartPartInfoProvider interface. The
// workspace will query for this interface when the view is shown.

[SmartPart]
public partial class MorticianView : UserControl,
    IMorticianView, ISmartPartInfoProvider
{

    // The workspace will call this method, passing the type of
    // SmartPartInfo that it knows how to use. This particular view
    // doesn't know how to provide any fancy type of SmartPartInfo,
    // so it simply provides a standard one containing title to
    // be shown on the tab. The Description property is never
    // used, so we simply pass null for it.


    public ISmartPartInfo GetSmartPartInfo(Type smartPartInfoType)
    {
        return new SmartPartInfo("Mortician", null);
    }
}
```

Chapter 4 Lab Exercises
Workspaces and Smart Parts

1. Generate a new SCSF project, or use the one from your previous example. Examine the workspaces present in the project.

2. Add a presenter and view to your project. In your controller, add the view to either of the workspaces that the project contains. Observe its behavior.

3. Add a label and a button to your view class. Add a click event for the button and a string property for the label to your view's interface. Connect them to your presenter.

4. Make your view support the *ISmartPartInfoProvider* interface, providing the *ISmartPartInfo* object to the workspace.

5. Add another view and presenter. Convert a *DeckWorkspace* to a *TabWorkspace*. Create several tabs programmatically and place an instance of this new view on each tab.

6. Write your own workspace class, deriving from the base class of your choice and implementing the *IWorkspace* interface. Switch the workspace in the project from a *DeckWorkspace* to your new workspace and observe its behavior.

Chapter 5
Shared User Interface Extension

A.   Problem Background

1. A composite user interface is, by definition, composed from disparate, independent pieces. Most Smart Parts displayed in workspaces provide their own controls for their specific input and output. However, users also expect to give commands and receive status notifications from the application's main user interface: the menu, toolbars, and status bars owned by the shell. Since a composite application is meant to be developed as a series of independent modules, the shell development team cannot possibly know about and deal with every user interface element that any of the modules might need to present to the user. Therefore, the CAB provides a mechanism for modules to independently modify the main user interface, within limits set by the shell developer.

You may find this problem familiar if you've been around the Windows development community for a decade or so. We first saw it back around 1993, when Microsoft was launching in-place activation in Object Linking and Embedding (OLE) 2.0. Many of the concepts used in extending the CAB user interface are likely familiar to you, although internally they're implemented in a completely different manner, being based on the .NET Framework instead of on COM and also being coupled more loosely than was OLE. But if seeing this stuff rings a muted bell in the dusty recesses of your memory, you're not the only one.

A sample CAB application demonstrating user interface extension is shown on the facing page in Figure 5-1 and is available in this book's sample code. It consists of a shell and a module containing a *WorkItem*. On the top level of the main menu, the File and Help dropdowns are owned by the shell, while the *WorkItem1* dropdown is owned by the *WorkItem*. The top toolbar is owned by the shell, as is the S button on it, while the W button is owned by the *WorkItem*. The lower toolbar, the one containing the A button, is a floating toolbar owned by the *WorkItem*, which is free to move within the shaded panel owned by the shell (try sliding it back and forth). On the status bar at the bottom of the window, I'll let you guess which of the two panels is owned by which party.

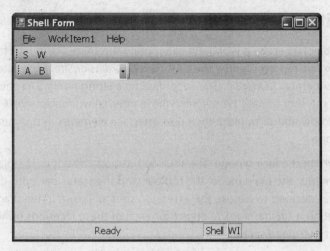

FIGURE 5-1 Sample CAB application with extended user interface. The shell owns all user interface elements except as otherwise labeled.

B.   Solution Architecture

1. The connection object whereby a *WorkItem* can modify the shared user interface is a *user interface extension site*, represented by the CAB class *UIExtensionSite*. Each extendible object requires a separate *UIExtensionSite*, identified by a string name. The shell developer decides, "I'll allow my subordinate *WorkItems*, whose internal workings I don't really know about or want to know about, to place their user interface elements in this location and this one and this one."

Figure 5-2 is shown on the facing page. The shell application developer has placed her own user interface elements, the main menu, the toolbar, and the status bar onto the shell application. She has decided to expose the extension sites as shown in the diagram. She does so by registering a *UIExtensionSite* object for each of these locations using the code shown on the next two-page spread.

Each subordinate *WorkItem*, for its part, contains code that queries for the presence of these extension sites and places additional user interface elements, such as menu items and toolbar buttons, onto them. The *WorkItem* in the lower-right corner has located the *UIExtensionSite* whose name is "StatusStrip" and is placing a panel onto it for showing the status information of the *WorkItem* to the user. The code for doing this is shown from page 8 on.

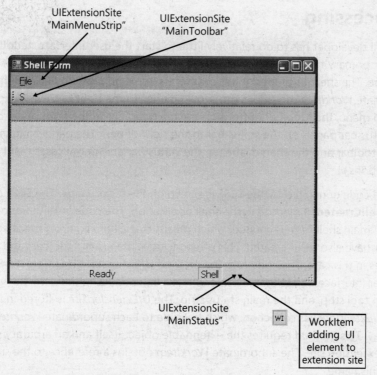

UIExtensionSite
"MainMenuStrip"

UIExtensionSite
"MainToolbar"

UIExtensionSite
"MainStatus"

WorkItem
adding UI
element to
extension site

FIGURE 5-2 Shell application showing UIExtensionSites.

C.   **Shell Processing**

1.   The shell developer has to do relatively little to start the user interface negotiation process. Her primary task is deciding which extension sites should exist and what their names should be. The shell developer of this chapter's sample application decided that each subordinate *WorkItem* can put its dropdown menus to the right of the File dropdown of the main menu, their toolbar buttons on the main toolbar to the right of her buttons, and their status bar panels on the status bar to the right of her. (The panel containing the floating toolbar and the shared label on the status bar are special cases that I deal with in later sections.)

The shell code generated by the SCSF is shown on the facing page. The SCSF overrides the **AfterShellCreated( )** method in the shell application. The code in this method is executed after the main shell form is created, which means that all the controls specified in the forms designer have also been created. This often includes menus and toolbars and status bars, as it does in this case. The code calls the method **UIExtensionSites.RegisterSite( )** for each user interface object that it wants to make extendable, in this case, the main menu, the main tool strip, and the main status strip. The *UIExtensionSite* is stored in the root work item's *UIExtensionSites* collection, which is visible to each subordinate *WorkItem* in the usual way. This method requires the extendable object itself and an arbitrary string name for the extension site. The subordinate *WorkItem* obtains a reference to the site by querying for this name.

CAB natively supports extending objects that derive from *ToolStripItem*, which includes just about anything in the *MenuStrip*, *ToolStrip*, or *StatusStrip* families. You can easily extend this support to other classes. I show an example at the end of this chapter (under an hour, 10 lines of code), so don't worry about it now. In this example, the shell programmer registers the main *ToolStrip* and the *StatusStrip* as extension sites.

```
protected override void AfterShellCreated()
{
    // delegate to the base class

    base.AfterShellCreated();

    // Register UI Extension sites for menu, toolstrip, status strip

    RootWorkItem.UIExtensionSites.RegisterSite (
      UIExtensionSiteNames.MainMenu,              // name of extension site
      this.Shell.MainMenuStrip);                  // object this site extends

    RootWorkItem.UIExtensionSites.RegisterSite(
      UIExtensionSiteNames.MainStatus,
      this.Shell.MainStatusStrip);

    RootWorkItem.UIExtensionSites.RegisterSite(
      UIExtensionSiteNames.MainToolbar,
      this.Shell.MainToolbarStrip);
}
```

D. Menu Processing

1. A *WorkItem* often tries to perform its user interface extensions as early in its lifetime as possible. The usual location for the extension code is the Run method of the *Controller* class on the *ControlledWorkItem*, which the *ModuleInit* class calls at startup time.

The facing page shows that method of the sample controller. For logistical convenience and clarity of explanation, I've broken down the user interface extension process into the separate utility functions that you see. Each extends one particular user interface object. Some of them are generated by the SCSF Wizard, whereas others I added myself.

Sometimes you just want to add menu items and toolbar buttons and receive commands from them. At other times, however, you want to modify these items after they've been added to the extended user interface, for example, changing the text of a StatusStrip panel or placing a checkmark in a menu item. In these cases, you need to save a reference to the user interface item at its creation time. I've added private member variables to the *Controller* class for holding these references.

```
public class MyOwnModuleController : WorkItemController
{
    // Private variables for holding references to user interface items
    // that need to be modified during the program's run

    private ToolStripMenuItem ToggleFloatingToolStripMenuItem;
    private ToolStripStatusLabel ToggleOnOffPanel;
    private ToolStrip ToolStripInTopPanel;

    // Run method called at startup time. Calls utility methods that
    // extend the various user interface elements

    public override void Run()
    {
        AddServices();
        ExtendMenu();
        ExtendStatusStrip();
        ExtendToolStrip();
        ExtendToolStripPanel();
        AddViews();
    }
    ... rest of MyOwnModuleController class
}
```

2. Adding menu items to an extension site provided by the shell isn't difficult. The facing page shows a code sample. First, we fetch the *UIExtensionSite* object that the shell has provided for us, accessing the *UIExtensionSites* collection of the *WorkItem* by means of the extension site's expected name (which the shell developer has to tell the *WorkItem* developer, probably out of band in some sort of functional specification).

Next, we create whatever menu items we want to add at the extension site. They appear in the order in which we place them onto the menu. The sample code shows a dropdown menu item containing a number of subitems, all assembled through hard-wired code. Your *WorkItem* could certainly assemble this dropdown menu in many other ways. For example, it could be assembled in the Visual Studio designer if you knew at design time which items it should contain, or through reading a configuration file or calling a web service if you wanted to dynamically fetch this information at runtime.

> **Note** One *ToolStripMenuItem* is saved in a member variable, whereas the rest are not. As we'll see when we look at the command handlers, responding to menu item clicks, turning menu items on or off, or even removing them from the menu can be done without further reference to the menu item object itself. But other actions, such as setting or removing a check from the menu item or changing its text, cannot be done in this manner; a reference to the original menu item is required for this purpose. So we save it in a member variable here when we create it, and we'll fetch it later when we need to use it.

The method **UIExtensionSite.Add ( )** hands the dropdown menu (with its internal collection of subitems) to the CAB framework, which places them onto the actual shell menu.

We've now placed the items onto the menu but haven't yet provided any sort of connection to code that would run when the user selects the menu item. The user can click these items until her head falls off, but they don't currently run any code, which is probably not a good design pattern. To connect handler methods to these menu items, we need to go to the Commands collection of the *WorkItem* and map them to menu items by means of the **AddInvoker( )** method. To this method, we pass the object to which we want the command wired (here it's the menu item) and name of the event that will trigger the command (here it's "Click"; see the event list of your particular object for available others). The string in the Commands indexer is the metadata identifier of the command handler method in the *Controller* class, so the CAB framework knows which method to call. To those command methods, we now turn to the next two-page spread.

```
private void ExtendMenu()
{
    // Fetch the main menu's extension site

    UIExtensionSite MainMenuSite =
        this.WorkItem.UIExtensionSites[UIExtensionSiteNames.MainMenu];

    // Create the new top-level menu item

    ToolStripMenuItem TopTi = new ToolStripMenuItem("WorkItem1");

    // Create the drop-down menu items

    ToolStripMenuItem ti1 = new ToolStripMenuItem("Do Something");
    ToolStripMenuItem ti2 = new ToolStripMenuItem("Enable/Disable Item Above");
    ToolStripMenuItem ti3 = new ToolStripMenuItem("Toggle Status Pane");
    ToolStripMenuItem ti4 = new ToolStripMenuItem("Show Moveable Toolbar in Panel");

    // Check the fourth one, and store it in the items collection
    // for further use, as we'll use it to toggle the movable toolbar.

    ti4.Checked = true;
    ToggleFloatingToolStripMenuItem = ti4;

    // Add it to the top level

    TopTi.DropDownItems.Add(ti1);
    TopTi.DropDownItems.Add(ti2);
    TopTi.DropDownItems.Add(ti3);
    TopTi.DropDownItems.Add(ti4);

    // Add the top-level item to the main menu using the extension site

    MainMenuSite.Add(TopTi);

    // Add the click command handlers

    this.WorkItem.Commands[CommandNames.Module1DoSomething].
        AddInvoker(ti1, "Click");
    this.WorkItem.Commands[CommandNames.Module1EnableDisableItemAbove].
        AddInvoker(ti2, "Click");
    this.WorkItem.Commands[CommandNames.Module1ToggleStatusPane].
        AddInvoker(ti3, "Click");
    this.WorkItem.Commands[CommandNames.Module1ShowMoveableToolbar].
        AddInvoker(ti4, "Click");
}
```

3. An excerpt from the *Controller* class of the *WorkItem* is shown on the facing page. It shows the command handler methods that are invoked by the CAB framework when the user selects a menu item or clicks a toolbar button (or whatever other events we place them on).

Each method is marked with the **CommandHandler** attribute, which marks it as the target for a command sent by the CAB framework. The text parameter inside the attribute specifies the name of the command to which it responds. These are the strings that you saw passed to the Commands collection when we called *AddInvoker* at the bottom of the preceding code sample.

The *WorkItem* maintains a collection of Command objects, indexed in the familiar CAB pattern by means of their string names. This allows you to view or modify the status of a Command object. The sample code for the second menu item handler enables or disables (turns it gray) the first menu item. The **CommandStatus** enumeration contains values for designating these states.

Note that no other information about the menu item, for example, its text or its checkmark status, is available from the Command object. If you want to modify these other properties, you have to save references to the menu items when you create them.

```
public class MyOwnModuleController : WorkItemController
{
    [CommandHandler(CommandNames.Module1DoSomething)]
    public void FirstMenuClickHandler(object sender, EventArgs e)
    {
        MessageBox.Show("Something Done");
    }
    [CommandHandler(CommandNames.Module1EnableDisableItemAbove)]
    public void SecondMenuClickHandler(object sender, EventArgs e)
    {

        // Check the first menu item and see if it's enabled. Turn it
        // on if not, off if so.

        if (WorkItem.Commands[CommandNames.Module1DoSomething].Status
                == CommandStatus.Enabled)
        {
            WorkItem.Commands[CommandNames.Module1DoSomething].Status =
                CommandStatus.Disabled;
        }
        else
        {
            WorkItem.Commands[CommandNames.Module1DoSomething].Status =
                CommandStatus.Enabled;
        }
    }

                < ... other handlers ... >

}
```

E.   StatusStrips

1. Extending the *StatusStrip* class works in essentially the same way as the previous *MenuStrip* example: you fetch the extension site, create the objects that go into it (*ToolStripStatusLabel* in this case), and add them with the site's *Add* method. The code is shown at the top of the facing page. You often don't hook up command handlers for *StatusStrip* panels because they are used primarily for display, but if you did, the procedure would work the same way as the *MenuStrip* example that precedes this one or the *ToolStrip* example that follows it.

   When I create the *ToolStripStatusLabel* object, I save it in a member variable for later use. When the user selects the Toggle Status Pane item from the menu, I run the command handler you see at the bottom of the opposite page, fetch the *ToolStripStatusLabel* from the controller's member variable, and change its text.

   The most common use of the status bar is a single shared piece of output on the left-most label. This is automatically handled in SCSF projects by the global CAB event named "StatusUpdate." I discuss this in Chapter 6.

```
private void ExtendStatusStrip ()
{

    // Fetch the StatusStrip extension site
    // placed here by the shell

    UIExtensionSite StatusStripSite =
      this.WorkItem.UIExtensionSites[
        UIExtensionSiteNames.MainStatus];

    // Create a new ToolStripStatusLabel, which is a rectangular
    // area that shows on the status bar

    ToolStripStatusLabel tssl = new ToolStripStatusLabel ("WI");
    tssl.BorderSides = ToolStripStatusLabelBorderSides.All;
    tssl.AutoSize = false;
    tssl.ToolTipText = "Work item status pane";
    tssl.AutoToolTip = true;

    // Use the ExtensionSite to add it to the status strip.

    StatusStripSite.Add(tssl);

    // Save the status label for use by menu item handlers
    // that will toggle it on and off.

    ToggleOnOffPanel = tssl;
}


[CommandHandler(CommandNames.Module1ToggleStatusPane)]

public void ThirdMenuClickHandler(object sender, EventArgs e)
{

    // Toggle the text in the panel to show we can control it.

    if (ToggleOnOffPanel.Text == "")
    {
        ToggleOnOffPanel.Text = "WI";
    }
    else
    {
        ToggleOnOffPanel.Text = "";
    }
}
```

F.   ToolStrips

1. A programmer may also want to add *ToolStripButtons* to any *ToolStrip* extension sites that the shell developer provides. This procedure works in the same manner as for MenuStrip items and *StatusStrip* items. This shouldn't surprise you because they share the same common base class of *ToolStripItem*. We fetch the UIExtensionSite from the *WorkItem* by means of its name, create the *ToolStripButton* we want to add, and tack them on using the method *UIExtensionSite.Add( )*.

In the case of a toolbar button, which exists to fire commands, we also want to add an invoker, which we do in the same way as we did for the menu example. In the sample code shown on the facing page, we wire the new toolbar button to the same handler function that the first menu item uses, a common technique.

Note that since we don't do anything with the *ToolStripButtons* other than sit there and receive commands from them, we don't bother saving references to them after placing them on the strip.

```
private void ExtendToolStrip()
{
        // Now fetch the toolstrip extension site

        UIExtensionSite ToolStripSite = this.WorkItem.UIExtensionSites[
                                        UIExtensionSiteNames.MainToolbar];

        // Create a button and add it

        ToolStripButton tsb = new ToolStripButton("W");
        ToolStripSite.Add(tsb);

        // Hook it up to the existing command handler for the first menu item.

        this.WorkItem.Commands[CommandNames.Module1DoSomething].
          AddInvoker(tsb, "Click");
}
```

G.   Non-ToolStripItem Classes: Writing a UIElementAdapter

1. The base implementation of the CAB contains classes for extending user interface sites deriving from *ToolStripItem*, but suppose we want to extend a different user interface site instead? For example, consider the case of floating, dockable *ToolStrips*. Users often expect to be able to move their *ToolStrips* around on the screen, perhaps hiding them and re-showing them later. What kind of stuff do we have to roll to get that working?

Creating floating, dockable *ToolStrips* turns out to be surprisingly easy and provides an instructive example that can help us with other classes as well. We create floating, dock-able *ToolStrips* by adding a *ToolStrip* to a *ToolStripPanel* on a Windows Form. The panel class handles operations such as docking. However, since this class does not derive from *ToolStripItem*, we can't register it as an extension site straight out of the box.

Each class that is to be extended needs to have a *UIElementAdapter* class written for it. This class knows the internals of the extended class well enough to be able to add and remove items from it. CAB comes with a *UIElementAdapter* for the *ToolStripItem* class from which the *MenuStrip, ToolStrip*, and *StatusStrip* all derive. You can examine it in the CAB source code. But it doesn't know how to add or remove items from the *ToolStripPanel* because the latter derives from *ContainerControl*. (If you look at the source code, you'll find its collection isn't called "Items" as it is in the *ToolStripItem* class; instead, it's called "Controls.") Therefore, I had to write my own implementation of *UIElementAdapter*, which is shown on the facing page. My class *UIElementAdapterToolStripPanel* derives from this base class. Its constructor requires the control for which it will add and remove items. It implements *Add* and *Remove* methods, which add and remove a ToolStrip from the *ToolStripPanel*. A production-quality implementation would probably check for null parameters and possibly check to make sure that a *WorkItem* removing a *ToolStrip* was indeed the one who originally added it. But these are small details.

The shell executes code to register an extension site using an object of this class, probably in the same location it registers all the other extension sites. In the sample program, the top *ToolStripPanel* is registered as an extension site, using one of my new adapters, thus:

```
// Register our private top ToolStripContainer panel

RootWorkItem.UIExtensionSites.RegisterSite(
    UIExtensionSiteNames.MainToolStripPanel,
        new UIElementAdapterToolStripPanel(
            Shell.toolStripContainer1.TopToolStripPanel));
```

Once we've registered the extension site, we can add items to and remove items from it in exactly the same way as we did for all the other user interface elements, as shown in the bottom method of the facing page.

If you use your *UIElementAdapter* class very often, you can make it part of the CAB framework by creating a class that implements the *IUIElementAdapterFactory* interface and registering it with the CAB catalog. Your class will then be recognized natively as the *ToolStripItem* classes are today. I'll leave that as an exercise for you.

```
class UIElementAdapterToolStripPanel : UIElementAdapter <ToolStrip>
{

    public UIElementAdapterToolStripPanel(ContainerControl cc)
    {
        this.cc = cc;
    }
    private ContainerControl cc;

    protected override ToolStrip Add(ToolStrip UIElement)
    {
        cc.Controls.Add (UIElement);
        return UIElement;
    }
    protected override void Remove(ToolStrip UIElement)
    {
        cc.Controls.Remove(UIElement);
    }
}

< controller class >
private void ExtendToolStripPanel ()

{

// Fetch the ToolStripPanel extension site

    UIExtensionSite ToolStripPanelSite = this.WorkItem.UIExtensionSites[
                                UIExtensionSiteNames.MainToolStripPanel];

// Create new toolstrip and button

    ToolStrip ts = new ToolStrip();
    ToolStripButton tsb = new ToolStripButton("A");
    ts.Items.Add(tsb);

// Save it in the WorkItem for later use

    ToolStripInTopPanel = ts;

// Add toolstrip to extension site panel

    ToolStripPanelSite.Add(ts);
}
```

H.   User Interface Modification Order

1. A *UIExtensionSite* doesn't allow extension items to be poked into any arbitrary location on the extended object that the shell developer provides. It only allows items to be appended, tacked onto the end of the object (the right end of the toolbar, the bottom of the popup menu, the right side of the main menu, and so on).

    The loose coupling of CAB means that different modules may be loaded in different use case scenarios. For the sake of having a consistent user interface, we need to ensure that different modules and work items extend the user interface in the correct order. If module A loads before module B in one scenario, and user interface extension is done automatically in the *Run* method, as was shown in the previous examples, then module A's menu items are shown to the left of module B's, as shown on the facing page in Figures 5-3. However, suppose a different use case leads module B to be loaded before module A. Now B's menu items are shown to the left of A's in Figure 5-4. The user will get very confused if his menu starts jumping around in front of his eyes. Even if it doesn't happen dynamically at runtime but only at load time in response to configuration settings, administration and documentation and training and testing will get blown apart if A's menu appears to the left of B's in some configurations and to the right of B's in others.

    There needs to be some way of enforcing standardization on the user interface. Since the *UIExtensionSite* allows this only by modifying the order in which the user interface items are extended, there must be some way of centrally enforcing this order. We could, of course, specify the load order of the modules. But there could easily be some reason not related to the user interface why modules needed to load in a certain order, perhaps related to security, and it would be better not to have the user interface dictate the load order. How can we solve this problem?

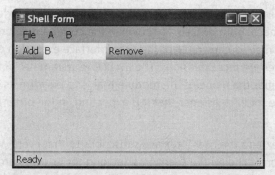

FIGURE 5-3 Application with menu items added in one order.

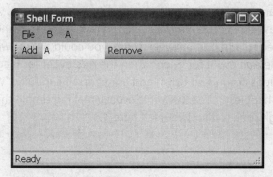

FIGURE 5-4 Same application with menu items added in reverse order.

2. The easiest way that I can think of to enforce a user interface modification order is through the loosely coupled event mechanism of CAB, as shown in the code sample on the facing page. The shell, which is the main orchestrator of the user interface extension dance, publishes an event called *AddUserInterfaceExtensions*. The *EventArgs* parameter of this event contains a string which specifies the name of the module that is to perform its extension. The shell fires the event once for each listener that it is expecting, in the order in which it wants the extensions to take place.

    Each module that wants to extend the user interface subscribes to that event. When it receives the event notification, it checks the *EventArgs*, looking for its own name. When the module finds its name, it modifies the user interface. The code is shown in the lower example on the facing page.

    If you really wanted to get fancy and do dynamic changes during operations, you could publish another event such as *RemoveUserInterfaceExtensions*. If modules A and B had their extensions up and you wanted to add C in between them, you could fire the *Remove* event containing B's name, then the *Add* event for C, and then the *Add* event for B again. But before you start whamming things around, sit down and ask yourself carefully, "What would make my users (not me) the happiest?" The answer is usually making simple things simple, and this might just be enough rope to hang yourself here.

```
< shell code ...>

private void toolStripButton1_Click(object sender, EventArgs e)
{
    // Fetch the event topic that tells the listening module to
    // add its user interface extensions.

    EventTopic et =
     RootWorkItem.EventTopics[
                      EventTopicNames.AddUserInterfaceExtensions];

    // Fire the event, passing the module identifier specified
    // by the user in the text box.

    et.Fire(this, new EventArgs<string>(this.toolStripTextBox1.Text),
        this._RootWorkItem, PublicationScope.Global);
}

< loaded module code ...
Event handlers that add or remove user interface extensions >

[EventSubscription(EventTopicNames.AddUserInterfaceExtensions,
   ThreadOption.UserInterface)]

public void OnAddUserInterfaceExtensions(object sender, EventArgs<string> eventArgs)
{
    // Check the module identifier to see if we're the addressee. If
    // so, then extend the menu (and possibly other UI elements).

     if (eventArgs.Data == "A")
     {
        ExtendMenu();
     }
}

[EventSubscription(EventTopicNames.RemoveUserInterfaceExtensions,
    ThreadOption.UserInterface)]
public void OnRemoveUserInterfaceExtensions(object sender,
    EventArgs<string> eventArgs)
{

    // Check the module identifier to see if we're the addressee. If
    // so, then remove our extensions to the menu.

     if (eventArgs.Data == "A")
     {
        UnExtendMenu();
     }
}
```

Chapter 5 Lab Exercise
User Interface Extension

1. Generate a new SCSF project. In your *Controller* class, add a menu popup and a few items, a toolbar button, and a status bar pane.

2. Figure out a way to add the Help menu item in its customary location at the rightmost end of the menu bar. Make it work properly if other popups are removed and added.

3. In your code that adds toolbar buttons, add controls that are not buttons to the toolbar, for example, a text box or a dropdown list.

4. Write a CAB service that fetches the list of items to add to the shared user interface from a web service. Store them in a config file and have the web service read them from that config file. See the ProfileCatalog web service in the Bank Teller reference application for ideas. If you really want to go to town, build this logic into a *Controller* base class from which you can derive all your *Controller* classes.

Chapter 6
Event System

A. Problem Background

1. The components of a CAB application sometimes need to announce asynchronous occurrences to each other. For example, the Smart Part that allows the user to select a customer might need to announce a new selection to any component that may be listening. ("Now hear this: whoever might be out there listening, the user has just selected the customer whose ID is X."). Other components in the CAB program might want to hear about that selection so that they can display their own business data keyed to the selected customer. ("Aha, customer X, eh? OK, let me fetch and display, in my own independent way, his current [mortgage, prescription, or whatever] records.")

   Accomplishing this task would be simple if the announcer and the listener(s) knew about each other at compile time, the way that forms know about the buttons that reside on them and announce user clicks using the .NET eventing mechanism. However, the whole point of the CAB architecture is that components are loosely coupled, not tightly coupled. We don't want them to have direct programmatic dependencies on each other. We want the Smart Part that allows the user to select a customer to announce the selection without knowing or caring which other components might be listening, thereby working properly with any listeners that might be developed or installed in the client application. Conversely, the listeners that take action when a new customer is selected, perhaps by fetching the newly selected customer's information from a database and displaying it to the user, don't want to be intimately tied to a particular piece of software that announces the selection, thereby working correctly with any code that does so.

   We need a general-purpose system that allows asynchronous occurrences to be announced, and the announcements to be received, in a loosely coupled way within a CAB application. We need the system to be easy to use, and we need its operation to be consistent with the other parts of the CAB framework.

> **Note** You will probably find the CAB event broker service conceptually similar to the COM+ event system described in my book *Understanding COM+* (Microsoft Press, 1999). The COM+ system also appears in the .NET Framework in the *EnterpriseServices* namespace. They are both loosely coupled, publish-and-subscribe, event systems. The main differences are (a) that the COM+ event system was intended to work primarily between applications, whereas the CAB event broker system works primarily within applications; and (b) the COM+ event system works internally by means of COM, whereas the CAB event broker system works internally by means of .NET.

B.   Solution Architecture

1. The CAB framework provides an event broker service that does a nice job of meeting the needs of both publishers of and subscribers to events. It does about what you think it ought to do, in more or less the way you think it ought to do it. But the nomenclature is a little confusing, so you need to follow this section carefully.

   An *event* is a thing, the occurrence of which can be announced. An *EventTopic* is the class of CAB object that contains the administrative information of a specific event, identified by a string name. Each *WorkItem* contains a collection of *EventTopic* objects, which percolate down to each descendent *WorkItem* in the usual CAB way.

   A *subscriber* is a method on an object that is called when an event is announced. Each *EventTopic* object maintains a list of subscribers to that particular event.

   To *fire* an event means to tell the *EventTopic* object to consult its list of subscribers to that event, and to notify each subscriber by calling the method specified in the *EventTopic*.

   The nomenclature confusion comes about with the ambiguous usage of the verb *publish* in the documentation, in the names of CAB objects, and in general developer conversation. It is sometimes used to mean "create an *EventTopic* object to which subscribers can subscribe," as in "Time-Warner publishes *Sports Illustrated* magazine." It is occasionally used as a synonym for *fire*, to announce the occurrence of an event to all registered subscribers, as in "Time-Warner published the swimsuit issue of *Sports Illustrated* last Tuesday." And sometimes it means "to connect a .NET tightly coupled event to a CAB loosely coupled event so that the firing of the former causes the firing of the latter." (This operation doesn't have a direct magazine equivalent. I discuss it later in this chapter.)

   The CAB event broker system is shown graphically in Figure 6-1:

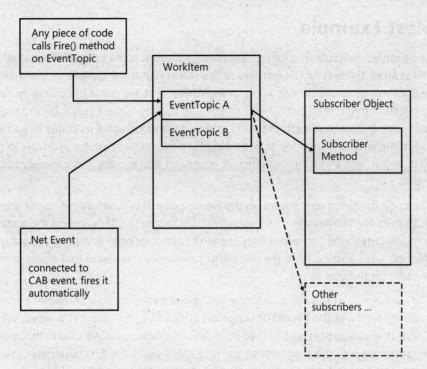

FIGURE 6-1 CAB event broker system.

C.   Simplest Example

1. The simplest example of a loosely coupled CAB event is the update of the shell form's main status label, the text on the left side of the status bar at the bottom of the shell form. An application generally contains only one of these, and any piece of code anywhere in the application might reasonably need to use it to convey current status information to the user. Rather than have each piece of code that wants to signal its status need to know the exact implementation of the status label (its name, its class, where and how to get a handle to it, and so on), a much more efficient approach is to expose it as a loosely coupled CAB event.

The code for doing so is shown on the facing page. It is automatically generated by the SCSF in its *ShellForm* class. The method StatusUpdateHandler contains the code that we want executed when someone fires the event announcing a new program status. The method takes a string from the *EventArgs* parameter and places that string into the status bar text for the user to see.

The connection of this method to the CAB event broker system takes place administratively by means of the **EventSubscription** attribute with which it is marked. When the CAB framework processes the shell form, this attribute tells CAB to add this method to the subscribers of the event whose name is specified by the first parameter, creating that *EventTopic* object if it doesn't already exist. As with other portions of SCSF, it is usually most convenient to keep these string names as constant values on a class to avoid collisions.

In this example, the main user interface thread creates the shell form and its status label. A rule of Windows Forms is that any Windows Forms object may be modified only from the thread that created it; therefore, the code in the subscriber method should be executed only from the main user interface (UI) thread. However, the code that fires the event might be running on a different thread. Since the subscriber doesn't know which thread the event is being fired from, it would normally have to check and marshal the call to the main UI thread if necessary. Fortunately, we can have the CAB framework do this for us automatically, by using the *ThreadOption* enumeration. In this case, the SCSF–generated code specifies *ThreadOption.UserInterface*. This selection tells CAB to always call this event method from the main user interface thread, regardless of the thread that fires the event, switching if necessary. Other options are *Background* (the subscriber receives the call on a thread other than the main UI thread, switching if necessary, to avoid tying up the main UI thread in a lengthy operation) or *Publisher* (the subscriber receives the call on the thread from which the publisher fires it, without any switch, for deterministic operation). Since CAB is used primarily for user interface operations, *ThreadOption.UserInterface* is most often the appropriate choice.

```
public partial class ShellForm : Form
{

    // This attribute tells CAB to subscribe this method to the event
    // whose name is specified by the first parameter, creating the
    // EventTopic object for it if necessary. The second parameter
    // tells CAB to always call this method from the main user
    // interface thread.

    [EventSubscription(EventTopicNames.StatusUpdate,
        ThreadOption.UserInterface)]

    public void StatusUpdateHandler(object sender, EventArgs<string> e)
    {
            // Take the status string that the caller passed us, and
            // place it into the status label for the user to see.

            statusLabel.Text = e.Data;
    }
}
```

2. Any piece of code that has access to an *EventTopic* object may fire an event. The most common way of obtaining an *EventTopic* is to index the *EventTopics* collection of the *WorkItem* by means of the event's string name, as shown in the first line of the code sample on the facing page.

Once you have the *EventTopic* object, you fire an event by means of the **Fire( )** method. You pass it the sending object and an *EventArgs*-derived parameter, just as for the usual .NET click events. The generic *EventArgs<>* class is generated by SCSF. It contains an element named *Data*, which has the type specified in the generic angle brackets, in this case string. You pass the contents of this *Data* element in the *EventArgs* constructor. This allows you to provide the subscriber with additional information about the event, such as the desired string to show in the status label. The generic types of the sender and the subscriber must match; otherwise an exception results.

The third and fourth parameters specify the scope of subscribers to which the event is announced, by using the **PublicationScope** enumeration. (Again, we see the nomenclature confusion; this usage of the p-word seems to imply a synonym for *fire*.) *PublicationScope.Global*, shown here, specifies that all subscribers in all work items are to be notified. If you wanted a narrower scope, you would specify a particular *WorkItem* in the third parameter (it's null in this example because we're firing globally), and a different choice of *PublicationScope*. Available options are *PublicationScope.WorkItem*, which specifies that only subscribers within that specified *WorkItem* will be called, and *PublicationScope.Descendants*, which means the specified *WorkItem* and all its children and subchildren. The whole point of CAB's loosely coupled event system is to facilitate communication between objects that don't contain strong references to each other. Objects in the same *WorkItem*, and sometimes even parent-child *WorkItem* relationships, are generally developed together in the same Visual Studio project, and therefore have easy access to strong references and the .NET tightly coupled event system. CAB's loose coupling of CAB is less beneficial in these cases. Therefore, *PublicationScope.Global* is the one that you see used the most often, by far.

Because the firing code doesn't know the subscribers' thread options, the firing code can't be sure that all the subscriber methods have returned or have even been called by the time the Fire method returns. Subscribers with *ThreadOption.Background* are called from the thread pool, and other thread confusion possibilities exist. Therefore, the firing code should be designed with one-way communication in mind, as pure "write-only" code.

```
private void toolStripButton1_Click(object sender, EventArgs e)
{
    // Fetch desired Event Topic object by means of its string name

    EventTopic et =
        _rootWorkItem.EventTopics[EventTopicNames.StatusUpdate];

    // Fire the event described by that EventTopic

    et.Fire(
                this,                              // sending object
        new EventArgs<string>(
          this.toolStripTextBox1.Text),           // event parameter
        null,                                      // sending work item
        PublicationScope.Global);                  // Publication scope
}
```

D.   More Complex Examples: Connecting .NET Events to CAB Events, and Programmatic Subscriptions

1. The tightly coupled .NET eventing system is used frequently throughout most .NET programs. It is often useful to be able to connect a tightly coupled .NET event to a loosely coupled CAB event so that firing the former also causes the firing of the latter.

   You can connect events declaratively, as shown on the facing page. The nonbold line of code beginning with *public event* is the declaration of an ordinary .NET event. The CAB attribute **EventPublication,** with which it is marked, tells the CAB Event Broker service, "Hey, I want you to subscribe to this .NET event, and when it is fired, I want you to fire this CAB event." You specify the string name of the CAB event and the publication scope as discussed previously.

   The lower paragraph of code on the facing page shows how this event is fired. Calling the .NET event by means of its name announces its occurrence to all tightly coupled .NET subscribers. Because of the *EventPublication* attribute, the CAB event broker service is one of these subscribers. When the event broker service receives the notification that the .NET event has been fired, it fires the CAB event named *MyOwnCABEvent* to all its loosely coupled CAB subscribers.

```
// This attribute specifies that we want the CAB event broker system
// to subscribe to the .NET tightly coupled event declared below. When
// anyone fires that .NET event, the CAB event broker will fire the CAB
// loosely coupled event named in the attribute, with the CAB
// publication scope specified in the attribute.

[EventPublication(EventTopicNames.MyOwnCABEvent,
  PublicationScope.Global)]

// This is the declaration of a standard .NET tightly coupled event.

public event EventHandler<EventArgs<string>> MyOwnDotNetEvent;

// User clicked the button. Fire the .NET event by calling it. This
// will cause the CAB event broker service to fire its loosely coupled
// CAB event, as specified in the EventPublication attribute above.

private void toolStripButton1_Click(object sender, EventArgs e)
{
  // Fire .NET event by means of its name

  MyOwnDotNetEvent (this,
      new EventArgs<string>(toolStripTextBox1.Text));
}
```

2. One of the more interesting things that you can do with this connection mechanism is to tie a .NET event from a control to a loosely coupled CAB event. For example, you can set up the system to automatically fire a CAB event when a button is clicked.

To do this, however, you need to manipulate the publication mechanism programmatically. You don't have access to the button's click event declaration to decorate it with the CAB *EventPublication* attribute. Therefore, you must fetch the desired *EventTopic* object and call its AddPublication( ) method, as shown on the facing page. You specify the object that fires the .NET event and the name of the .NET event. CAB uses the reflection API internally to add a listener for that .NET event. You must specify the work item and publication scope with which to fire the CAB event when the .NET event is fired. Interestingly (OK, I'm a geek), the *WorkItem* parameter cannot be null here even if the publication scope is global.

Note The controls fire their events passing the object *System.EventArgs* as their parameter, not the generic version seen previously in CAB events. If you want your CAB event handler to work properly, you have to set it to accept this version, as shown on the bottom of the facing page.

```
// User clicked the Set Up Publication button.

private void toolStripButton4_Click(object sender, EventArgs e)
{

    // Fetch the EventTopic object that describes the named event.

    EventTopic et =
        _rootWorkItem.EventTopics[EventTopicNames.ButtonClickCAB];

    // Tell the CAB event broker service to subscribe to the .NET event
    // named "Click" on the specified button object. When that .NET
    // event notification is received, fire the CAB event described
    // by the EventTopic object, with global scope.

    et.AddPublication(
        this.toolStripButton5,       // Object that fires .NET event
        "Click",                     // Name of .NET event
        _rootWorkItem,               // Work item to fire CAB event
        PublicationScope.Global);    // Scope of CAB event
}

// This method subscribes to the CAB event that is tied to the
// specified .NET event in the above code sample. Note the use of
// System.EventArgs as the second parameter, as this is what the button
// that fires the .NET event passes.

[EventSubscription(EventTopicNames.ButtonClickCAB,
    ThreadOption.UserInterface)]

public void ButtonClickCABEventHandler (object sender, EventArgs e)
{
    MessageBox.Show(e.ToString());
}
```

3. To complete the discussion of the CAB event broker system, I point out that CAB event subscriptions can also be added programmatically, as shown on the facing page. You fetch the *EventTopic* object controlling the event to which you want to subscribe. You then call its AddSubscription( ) method, specifying the object and method name on which you want to receive subscription calls. You also have to specify the *WorkItem* that owns the subscribing object, for use in publication scope determinations, and the thread option with which you want the subscriber method called.

```csharp
// User clicked this button. Look at the event topic
// and hook up to it programmatically

private void toolStripButton2_Click(object sender, EventArgs e)
{
        // Find the event topic with the name of the one
        // that we want

    EventTopic et =
        _rootWorkItem.EventTopics[EventTopicNames.StatusUpdate];

        // Add a subscription, calling the specified method on
        // this object when the event is fired

    if (et != null)
    {
        et.AddSubscription(this, "ProgrammaticHandler",
            _rootWorkItem, ThreadOption.UserInterface);
    }
}


// This is the method that we're subscribing.

public void ProgrammaticHandler(object sender, EventArgs<string> e)
{
    MessageBox.Show("Programmatic handler" + e.Data );
}
```

Chapter 6 Lab Exercises
Event System

1. Generate an SCSF project. Add an event publication using the SCSF Wizard, with the name of your choice. Add an event subscription, popping up a message box so that you know that you've hit it.

2. Add two methods, programmatically adding a handler for an event subscription and programmatically firing it.

3. Add a list box to the shell form. Read the *EventTopics* list from the root work item and show the events in the list box.

4. Write code to enable you to select an event from the list, hook a handler to it, and spy on it.

Chapter 7
Action Catalog Service

A.   Problem Background

1.   The benefits of loose coupling are not confined to the display of visible controls on user interfaces. A common problem in composite applications is that of coordinating loosely coupled parts of business logic. One work item wants to take a certain action but doesn't know if doing so would be OK with the work items in the composite application, especially since the initiator doesn't know and doesn't really want to know exactly which other items are currently loaded in the application. It wants to say something like, "Hey, I'm about to do [this thing]; if anyone objects, let him speak now or forever hold his peace."

Consider the hospital application that we've been working with throughout this book. Switching the selected patient when the user clicks a new patient in the selection box is a common and important task, which is and should be done frequently. However, the user may have done a lot of work on one of the tabs (say, entering a complex pharmacy regime) and not yet saved it. Switching the selected patient would cause all that work to be lost. We'd like a way for the each *WorkItem* on the various tabs to be notified of the switch and to prevent it if desired. The pharmacy tab might, for example, pop up a message box saying, "Save patient data, discard, or cancel switching the current patient?" In the first two cases, it would allow the switch to continue (after saving or not saving the pharmacy data). In the last case, it would abort the switch.

This may sound like a job for the eventing system, but its design doesn't fit this case well for several reasons. First, there's no good way for event methods to return output so as to signal the veto. Second, because of possible threading option choices of event methods, the sender can't be sure when an event has been received by all parties. You could hammer together a system on top of the event broker that would address these problems, but it would require writing a lot of infrastructural code, something that we always try to avoid.

We would like a mechanism that would allow us to

a.   call a method by means of a string name, without needing to know where it lives or who implements it;

b.   definitively get output data from this method regardless of threading options;

c.   allow other work items in the composite application to be informed of the method call prior to its invocation and to have the ability to veto it; and

d.   accomplish all these tasks in a loosely coupled manner consistent with the rest of CAB.

B.   Solution Architecture: The Action Catalog Service

1. The SCSF provides just such a capability in its *action catalog service*. Its operation is shown in Figure 7-1 on the facing page.

A conditionally executed method is known as an *action*. A module registers its actions with the action catalog service, either declaratively or programmatically, most often at the time that the module is loaded.

A module wishing to be notified of the requested invocation of an action, and to be allowed to inspect it and optionally veto it, provides the action catalog service with an object called an *action condition*. This object contains a Boolean method named CanExecute. An action condition may be registered for one named action (a *specific condition*) or all actions in the catalog (a *general condition*).

When a client wants to invoke an action, it calls the action catalog service's Execute() method, passing the name of the action that it wants to invoke and the parameters to be passed to it. The service then invokes the CanExecute method of all action conditions, specific and general, that apply to that action. If all conditions return the value *true*, then the service invokes the action method. If any of them return *false*, then it doesn't. This process is conceptually similar to resource managers voting on the outcome of a transaction, commit or abort.

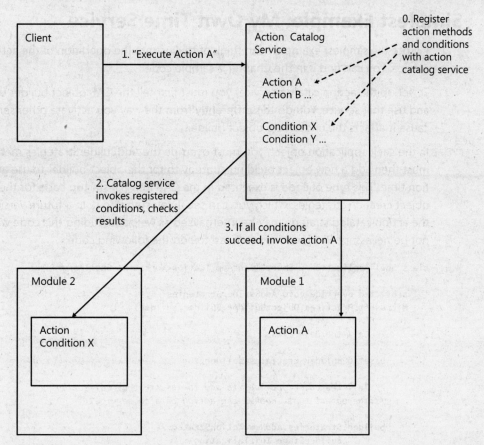

FIGURE 7-1 Schematic diagram of the action catalog system.

C.   Simplest Example: My Own Time Service

1. I wrote the simplest example I can think of to illustrate the operation of the action catalog service. You can find it in the Chapter 7 sample code.

To set up the action catalog service, you must first tell the CAB object builder to activate and use that service. You do this differently from the way you activate other services because it affects the underlying object builder.

In the shell application object, you must override the AddBuilderStrategies method. You must then add a new object building strategy to for the object builder to use at initialization time. This type of code is executed in the CAB internal startup code for the other object creation strategies, which you can see if you step into it. In a future version of CAB, the action catalog strategy may be internalized as well, and adding this code will therefore not be necessary. But for now, you need to add the following code:

```
class ShellApplication : SmartClientApplication<WorkItem, ShellForm>
{
    protected override void AddBuilderStrategies(
    Microsoft.Practices.ObjectBuilder.Builder builder)
    {
        // Call the base class

        base.AddBuilderStrategies(builder);

        // Tell the object builder to add to its strategy collection
        // the object that invokes the action catalog service

        builder.Strategies.AddNew<ActionStrategy>(
                BuilderStage.Initialization);
    }
}
```

2. Having turned on the action catalog service, you must register the methods that you want the action catalog service to be able to invoke. The easiest way to do this is declaratively using program attributes, in a manner conceptually similar to event subscriptions. Your action methods can live anywhere, but you will probably find it most convenient to place them on your work item controller class. Each method that you want to be invoked via the action catalog service must be decorated with the **[Action]** attribute, specifying the string name by which you want it listed in the catalog. As with the other shared string names in SCSF, you probably want to keep track of these names in a separate enumeration so as to avoid collisions. The action methods are placed into the catalog when the class that contains them undergoes its CAB framework processing. Alternatively, you can register action methods programmatically via the action catalog method RegisterActionImplementation (not shown). You can remove an action method from the catalog via the method Remove ActionImplementation (not shown either).

The parameter list of the action method must always be exactly as shown in the following sample: two objects, the first of which is the original caller of the action, the second of which is the parameter passed to the action by the caller. Thus:

```
// Class containing all Action Catalog Service
// actions in this module

public class Controller
{
    // Each action needs to be marked with this attribute
    // AND have this specific parameter list

    [Action(ActionNames.ActionSayHello)]

    public void ActionSayHello(object caller, object target)
    {
        MessageBox.Show("ActionSayHello called");
    }
}
```

3. The facing page shows the code for invoking an action. In this example, I chose to invoke the action from the shell form, in response to the user clicking a toolbar button. I've set up service dependencies in the shell form so that the CAB framework injects it with the root work item and the action catalog service.

You invoke an action by calling the method **IActionCatalogService.Execute( )**. You pass it the name of the action you want to invoke. The next two parameters are the work item in which the caller resides (the caller probably doesn't know or care about the work item in which the action method resides) and the caller object itself, so the action conditions can know who triggered the call which they have the opportunity to veto. You can also pass it an optional parameter containing data that it needs to do its work. Here, I simply pass null.

This code causes the catalog service to run all the action conditions registered for this action and invoke the action if they all succeed. The Execute method's return type is *void*, so the caller doesn't know whether the action succeeded unless you write code to somehow collect this information. A later example in this chapter shows one possible technique for doing this.

If you want to find out in advance whether the action conditions would allow an Execute request to be successful, you can check via the method IActionCatalogService.CanExecute (not shown). Using this method causes the catalog service to invoke the registered action conditions but NOT to invoke the action method, even if they all succeed. It returns a Boolean telling you whether all the conditions allowed the action. Keep in mind that this result is advisory, not concrete. The action conditions might change their minds between the time you get the result back and the time you invoke the action.

```csharp
public partial class ShellForm : Form
{

    // Service dependencies. We need the root work item and
    // the action catalog service

    private IActionCatalogService _actionCatalogService;
    private WorkItem _rootWorkItem;

    [ServiceDependency]
    public WorkItem RootWorkItem
    {
        set { _rootWorkItem = value; }
    }

    [ServiceDependency]
    public IActionCatalogService ActionCatalogService
    {
        set { _actionCatalogService  = value; }
    }

    // User clicked the "Execute Action" button
    // Call the SayHello method using the action catalog

    private void toolStripButton1_Click(object sender, EventArgs e)
    {
        _actionCatalogService.Execute(
            ActionSayHello, // action name
            rootWorkItem,   // context work item
            this,           // caller object
            null);          // optional parameter
    }
}
```

4. Providing an action condition is quite simple. You write a class that implements the **IActionCondition** interface. This interface contains one method, **CanExecute,** that accepts all the parameters passed by the original caller to the Execute call. Your action condition looks at its own internal state, maybe looks at these parameters, looks at anything else it cares about, and decides whether the call should be allowed to proceed. It returns *true* if so and *false* if not. The code is shown here:

```
public class MyOwnActionCondition : IActionCondition
{
    #region IActionCondition Members
    // To simulate an action condition, pop up a message box asking
    // the user whether to allow the call to proceed or not. If the
    // user clicks "Yes", return true, thereby allowing the action
    // method invocation. Otherwise, return false, thereby causing the
    // action invocation to be cancelled.

    public bool CanExecute(string action, WorkItem context,
        object caller, object target)
    {
        if (MessageBox.Show("Well?", "Allow action ?",
            MessageBoxButtons.YesNo) == DialogResult.Yes)

        {
            return true;
        }
        else
        {
            return false;
        }
    }
    #endregion
}
```

Note The first action condition to return *false* dooms the action invocation to eventually abort because the result cannot be changed once it is returned. You might therefore think that the action catalog service would then stop checking the rest of the action conditions because their return values have become irrelevant to the overall question of whether the action method will be called. For whatever reason, the action catalog does not currently do this. It continues to call all conditions registered in the catalog for the specified action, even after one of them returns *false*. There may or may not be a good logical reason for this, but that's the behavior that we have. If an action condition really needs to abort the process, it should throw an exception, which the client code would catch.

5. You cannot currently register your action conditions declaratively, as you can your action methods. Instead, you must make programmatic calls to the action catalog service. You will often find this approach most convenient in the AddServices method of a *ModuleInit*, as shown in the following code sample, or perhaps a controller's Run method. You first fetch the action catalog service, which can be done via explicit call (as shown here) or via dependency injection if desired. You then call the method **RegisterGeneralCondition( )** to add your action condition to all action methods in the catalog or RegisterSpecificCondition() to add it to a particular named action method. The methods RemoveGeneralCondition() and RemoveSpecificCondition() (not shown) undo these operations. Thus:

```
public class Module : ModuleInit
{
    private WorkItem _rootWorkItem;

    public override void AddServices()
    {
        // Delegate to the base class

        base.AddServices();

        // Fetch the action catalog service from the
        // root work item

        IActionCatalogService catalog =
                _rootWorkItem.Services.Get<IActionCatalogService>();

        // Register the action condition as a general condition,
        // that is, one that applies to all actions

        catalog.RegisterGeneralCondition (new MyOwnActionCondition());
    }
}
```

D.   More Complex Example: Passing and Modifying Parameters

1. It is reasonable and likely that an action method would like some sort of input parameter to help it do its job. You can pass this as the last parameter to the Execute method. It can be an object of any type. In the hospital example, you might expect that this would be the new patient ID. Counterintuitively, we don't do that. The reason is that we want the currently selected patient ID to come from one and only one place, that being the clinical context service. That way, it can never get out of sync.

You can also use the object parameter as a means to get output from the action method and even the action conditions. The parameter is passed by reference, so the action conditions can modify it. This technique we do use in the hospital example. I've defined an object class called *ClinicalContextChangeParam*, shown in the following code sample. It contains a Boolean parameter that an action condition sets to *false* to indicate to the caller that it has vetoed the action. (Since the method IActionCatalog.Execute has a void return, there is no other good way for the caller to get this information.) I've also provided a list of strings that action conditions can use to provide more information to the caller about why they might have vetoed the action. The code is shown on the facing page.

```
public class ClinicalContextChangeParam
{
    public bool ChangeOK ;
    public List <string> VetoList;

    public ClinicalContextChangeParam()
    {
        ChangeOK = true;
        VetoList = new List<string>();
    }
}
```

```
public class PharmacyActionCondition : IActionCondition
{
    public bool CanExecute(string action, WorkItem context,
        object caller, object target)
    {
        // Show user the dialog box explaining the choice

        DialogResult dr = MessageBox.Show(
        "You have done a large amount of work on"
        "the pharmacy" tab. Do you want to save it?",
            "About to switch patient", MessageBoxButtons.YesNoCancel);

        // If user clicks "Yes", then save the current data and
        // return true to allow the switch

        if (dr == DialogResult.Yes)
        {
            _MyOwnSave();
            return true;
        }

        // If the user clicks "No", then allow the current data to
        // be discarded and return true to allow the switch

        else if (dr == DialogResult.No)
        { return true; }

        // User clicked "Cancel", meaning to cancel the switch.

        else
        {
            // Cast the parameter into ClinicalContextChangeParam

            ClinicalContextChangeParam cccp =
                    (ClinicalContextChangeParam) target;

            // Set its ChangeOK Data member to false, advising the
            // caller that the action has been vetoed.

            cccp.ChangeOK = false ;

            // and add our name to the provided veto list, in case
            // the caller cares who has vetoed the action

            cccp.VetoList.Add("PharmacyModule");

            // The above changes are advisory only. Returning
            // false here is what actually vetoes the action.

                return false;
        }
    }
    return true ;
}
```

Chapter 7 Lab Exercise
Action Catalog Service

1. Generate a new SCSF project. Modify the builder strategy in the shell application to use the action condition strategy.

2. Add an action class to your module. Add a method to this class, decorating it with the attribute that marks it as an action. Place code in the method that allows you to detect its being called. Make sure that you follow the correct method signature; otherwise, you get a confusing error message. In your controller class, register this action class.

3. From some convenient location in your shell application, execute the action. If you like, check first with the CanExecute method.

4. Create an action condition class that allows you to accept or reject an action. Register it with the action catalog service. Execute your action again and observe its success and failure.

5. Experiment with passing parameters from the original caller, through the action condition, to the action method. Experiment with modifying the parameter in the action condition and also the action method.

Chapter 8
CAB and WPF

A.   Problem Background

1. The original releases of CAB and SCSF used Windows Forms (WF) for their graphical presentation layers. However, many programmers today are looking to migrate from Windows Forms to Windows Presentation Framework (WPF). There are a variety of reasons for this: better performance, a richer toolkit, separation of code from appearance using XAML. We need to figure out a way to combine this richer presentation layer with the benefits of loose coupling that we've seen with CAB and SCSF.

   Many parts of CAB won't need to change at all. For example, there's no reason that dependency injection, or the event system, or the action catalog, or the module loader and enumerator services, should depend on Windows Forms and hence need modification. And as you'll see, they don't. The CAB design team did a good job of minimizing dependencies on Windows Forms, and of placing the irremovable dependencies into small and well-circumscribed locations so that anyone needing to modify them can locate and change them without too much trouble.

B.   Official Solution Architecture: Interoperation

1. The May 2007 release of the SCSF contained Microsoft's first official attempt at support-
    ing WPF in CAB applications. You can see it most easily by examining the WPF Integration
    Quickstart sample program that comes with this release, a screen shot of which is shown
    below. The goals of this WPF support package are:

    a) Providing an easy way to use WPF controls as SmartParts in CAB applications, while
        interoperating seamlessly with WF controls in the same application.

    b) Providing official guidance about Microsoft's recommended mechanisms for
        incorporating WPF functionality into CAB applications today.

    c) Providing a present development path that will lead to the easiest upgrade in
        future implementations of CAB and related technologies.

> **Note** WPF provides many powerful graphical display features to which programmers have
> not had easy access in the past. It is common for programmers to sprinkle these new features
> over their UI designs like some magic fairy dust, foolishly assuming that since it's new and
> powerful, it must somehow be a Good Thing which cannot fail to enhance their applications.
> The screen shot in Figure 8-1 clearly demonstrates this attitude, but it's wrong! The reflections
> of the organizational chart buttons, for example, do not enhance the user's experience in any
> way. On the contrary, they distract the user by providing twice as much text to decipher, half
> of which is redundant and hard to read because it's upside down. The color gradients on the
> buttons, likewise, add no meaning and only distract the user's attention. The designers of this
> sample probably only wanted to demonstrate that they were, in fact, using WPF, and they
> accomplished this. But in a production application, distracting the user with extraneous
> nonsense because a UI designer doesn't know any better constitutes malpractice. Unless your
> application provides sex or sex equivalent, your user almost certainly wishes he were
> somewhere else, doing something else, instead of using your application. He doesn't want
> "cool," he wants "finished." When you start using WPF, ensure that you do so for good and
> not for evil, lest you be cast into the extra-hot fires that hell reserves for bad UI designers who
> please themselves instead of their users.

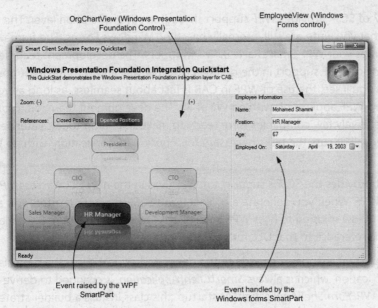

FIGURE 8-1 WPF integration Quickstart sample program.

2. The May 2007 of SCSF provides WPF support through an interoperation layer. The CAB applications that you write are still internally based on Windows Forms. The framework automatically wraps your WPF Smart Parts in WF wrappers. The development team felt that this would provide WPF support in the quickest possible way and with the least possible disruption. As you'll see, the use of WPF in CAB is just about seamless, as long as you don't insist on a pure WPF application with no WF at all. The development team felt, probably rightly, that relatively few companies were ready to make that jump today. Later sections of this chapter will discuss the tasks necessary for a pure WPF application, with no WF whatsoever.

The code that provides the SCSF's support for WPF lives in the new DLL *Microsoft.Practices.CompositeUI.WPF*. When you generate a project using SCSF, the wizard offers you a checkbox that says "Allow solution to host WPF Smart Parts," as shown in Figure 8-2. Selecting this box adds a reference to that DLL for your new project.

Two other important changes are made to the generated project. First, the base class of your shell application, which is always *SmartClientApplication*, is changed to derive from the new class *WPFFormShellApplication*. At startup, this class installs a builder strategy that knows how to create Smart Parts based on WPF controls. It also adds a service that knows how to wrap up WPF controls in a Windows Forms wrapper and extract them again. These changes happen at low levels and aren't often used directly by application programmers.

The main visible change is that the workspaces that are placed onto the shell form change from the regular CAB workspaces to the new WPF interoperation workspaces, such as *Microsoft.Practices.CompositeUI.WPF.DeckWorkspace*. We will now turn our attention to those workspaces.

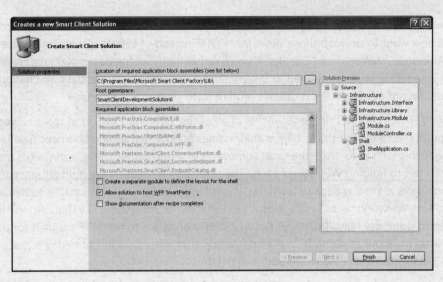

FIGURE 8-2 Add to your new project a reference to the DLL.

3. A workspace, as I'm sure you remember, is a container for Smart Parts (views, controls). We now want to develop these views from WPF instead of WF. If you remember the *IWorkspace* interface from Chapter 4, you'll remember that the parameters that represented Smart Parts were all defined as *System.Object*, the universal base class. So changing the base class of our Smart Parts doesn't require us to change this interface.

The WPF DLL contains new equivalents of all of the WF-based CAB workspaces. Even though they have the string WPF in their names and in their toolbox bitmaps (see facing page), this is (as is so often the case) somewhat of a misnomer. These workspaces are not restricted to holding WPF Smart Parts. They can, and frequently do, hold either type. You don't have to set any flags or anything, it just happens. So they really ought to be called "WPF or WF workspaces," or "agnostic workspaces," or something like that. When I first started using the May 2007 SCSF, I used a WPF workspace to hold WF controls for a couple of weeks without realizing it. There's no reason that these couldn't someday replace the CAB workspaces, as they swing both ways.

You need to add these workspaces to your Visual Studio toolbox. You do this with the Choose Items dialog box in the usual way. Just make sure that you select the WPF DLL, which you'll know you've done when you see the icons as in Figure 8-3.

The workspace's implementations are very similar to the original WF equivalents. If you look at the code, you'll see that the primary difference is a new composer class called *ElementHostWorkspaceComposer*. Internally it uses a wrapper from the *System.Windows. Forms.Integration* namespace to place around WPF controls. You can see a code sample on the facing page of the read-only property *ActiveSmartPart*. The code looks at the value that the workspace is maintaining internally for the active Smart Part. If it's of the WPF wrapper class, then it unwraps and returns it. Otherwise it just returns it, since unwrapping is unnecessary. You don't really have to worry about this in your applications, it just works seamlessly in the way that you're expecting it to. But it's always good to have at least some notion as to what's going on under the hood.

```
// Method implemented by a ElementHostWorkspaceComposer

public object ActiveSmartPart
{
  get
  {
      // If the active Smart Part is a wrapped WPF control, then unwrap
      // it and return the control.

      if (activeSmartPart is ElementHost)
      {
        return elementHosts.Unwrap (activeSmartPart as ElementHost);
      }

      // Otherwise it must be a regular WF control, so just return it.

      return activeSmartPart;
  }
}
```

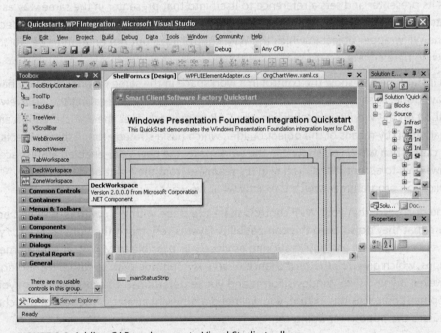

FIGURE 8-3 Adding CAB workspaces to Visual Studio toolbox.

4. The main thing that we want to do with WPF is to develop views composed of WPF controls. Adding a WPF view to our SCSF project is easy. We simply right-click on the project as before. Now the SCSF context menu contains the choice "Add WPF View with Presenter," just below the original WF-based "Add View with Presenter." You select the former option and it generates the usual control, interface, and view classes that we encountered in Chapter 4.

The presenter and the interface don't change at all now that we're using WPF. That's one advantage of the model-view-presenter architecture that I discussed earlier. Since the presenter only knows the view through its interface, the presenter doesn't know or care that the implementation of that interface is now being provided by a WPF control rather than a WF control.

The one thing that does change, modestly, is the control class itself. You'll see that it is now a WPF class, deriving from **System.Windows.UserControl** (top code sample). You'll see that it has a .XAML file containing its layout markup (middle code sample). The control creates its presenter and sets a reference to itself into that presenter in the same way as a WF control, so I won't bother showing it to you. The Dispose method changes a little because of the WF wrapper object in which the WPF control lives, but not enough to be worth showing here either.

You instantiate and show the control in a workspace in the same manner that you have always done, as shown in the bottom code sample. Since you are calling the *AddNew* method on the *WorkItem's* collection, the CAB framework does the actual creation and initialization by means of the object builder, which has a strategy for creating objects of this class. Then when you show it in the workspace, the new WPF-enabled workspace knows how to accept it and display it. It's seamless from the point of view of the application programmer, which is exactly what we want.

If you don't care that your WPF controls share the stage with WF controls, or if you actively desire that they do so, then this compatibility layer is very much the right choice for you. It's very easy to use, and you're sticking as closely as possible to Microsoft's recommendations, which give you the best chance of an easy upgrade path to future technologies. However, if you dislike Windows Forms and you're determined to expunge it from your development practices this instant, you will need a different strategy. The rest of this chapter discusses the options for doing so.

```
using System.Windows;
using System.Windows.Controls;

[SmartPart]
public partial class OrgChartView : UserControl, IOrgChartView,
 IDisposable
{

}
```

Listing N, WPF-based SmartPart class

```
<UserControl
    x:Class="Microsoft.Practices.QuickStarts.WPFIntegration.OrganizationChart.OrgChartView"
    xmlns="http://schemas.microsoft.com/winfx/2006/xaml/presentation"
    xmlns:x="http://schemas.microsoft.com/winfx/2006/xaml"
    xmlns:mc="http://schemas.openxmlformats.org/markup-compatibility/2006"
  >

</UserControl>
```

Listing N+1, WPF-based SmartPart XAML file

```
private void AddViews()
{
  // Instantiate a new WPF view, same as the old WF way

    OrgChartView view =
        WorkItem.SmartParts.AddNew<OrgChartView>("OrgChartView");

    // Show it in a workspace, ditto

  WorkItem.Workspaces[WorkspaceNames.LeftWorkspace].Show(view);
}
```

Listing N+2, instantiating and showing a WPF-based SmartPart in a workspace.

C. Solution Architecture: Porting the CAB Libraries

1. The CAB libraries need to be converted to use WPF as their presentation layer. This task is not conceptually difficult but takes some amount of time and knowledge of both WPF and CAB. I would call the changes *evolutionary* rather than *revolutionary*.

This chapter discusses the port of the CAB libraries produced by Kent Boogaart, a freelance programmer living in South Australia. It is currently available under an open source license at *http://www.codeplex.com/wpfcab*. Corporate developers who don't want or aren't allowed to depend on open source projects can reasonably study it and perform the migration themselves.

CAB, as you've seen, comes in three dynamic linked libraries (DLLs): ObjectBuilder, Composite UI, and *CompositeUI.WinForms*. The first two need no modifications at all. Boogaart made a couple enhancements, but they aren't necessary, and if you object to them, you can remove them without hurting anything. In CompositeUI, he added the SmartPartActivating method to the *IWorkspace* interface, to be called before a SmartPart is activated rather than after. The *IComposableWorkspace*, which derives from *IWorkspace*, contains a new RaiseSmartPartActivating method to support it. The classes *Workspace* and *WorkspaceComposer* that implement these interfaces contain code to support the new methods. These are the only changes to those two DLLs.

Using a file comparison tool such as WinDiff, you can compare the original CAB source files with the ones that Boogaart produced. I won't describe them in detail, but broadly speaking, here are the sorts of classes to which he made modifications:

- Application and startup classes to accommodate the different startup sequence of a WPF application

- Builder strategies to accommodate different construction rules of WPF objects

- Workspaces to accommodate WPF-based controls and WPF-based Smart Parts

- *UIElementAdapter* and related classes to support WPF menus, toolbars, and status bars; also *CommandAdapter* classes to receive commands from these WPF objects. .

- Display code of the Visualizer object, for displaying CAB items in a WPF-based spy program

D.   Further Solution Architecture: Porting the SCSF

1. I have found that CAB is necessary but not sufficient for developing applications. Using CAB cost-effectively requires you to use SCSF, as using the Microsoft Foundation Classes cost-effectively required the use of Class Wizard. So porting CAB to WPF was absolutely necessary and wonderful and good, but it's not sufficient for developing industrial applications.

I therefore decided to convert the SCSF to use WPF and did so with about two weeks' worth of effort. The guidance package containing my port comes with the code download for this book. You install it into Visual Studio with the Guidance Package manager, as explained in the sample code instructions. Then, when you open Visual Studio and select File | New Project from the main menu, you see the project types shown in Figure 8-4. I've provided templates for both a classic WPF desktop application and browser-hosted XAML Browser Application (XBAP) application. Thus:

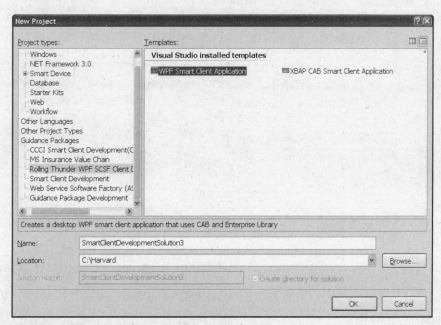

FIGURE 8-4 Using My WPF SCSF guidance package.

2. I would describe porting the SCSF to WPF as tedious but not difficult. I went to the SCSF help file, read the instructions about cloning the guidance package, and followed them. Then I just worked on the projects one at a time: removing the reference to *System.Windows.Forms* from the Interface library project, seeing what complained, fixing it, repeating as necessary until done. Then I moved on to the Library project and so on. A sample screen shot of guidance package development is shown Figure 8-5:

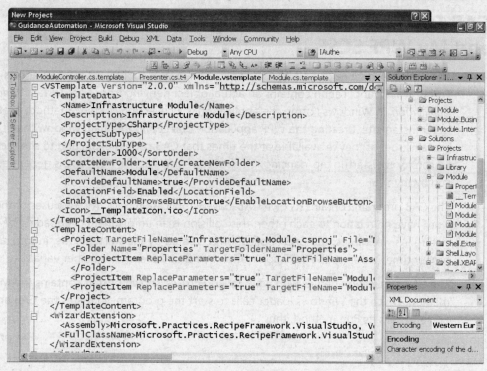

FIGURE 8-5 Some of the internals of my WPF SCSF guidance package.

E.   CAB Example Based on Boogaart's WPF CAB DLLs and My WPF SCSF

1. The startup sequence of a CAB/SCSF application in WPF is different from Windows Forms. A WPF application contains an actual object deriving from class *System.Windows. Application*. In Windows Forms, the application is a collection of static methods that you never instantiate. Creating the WPF application object and the main window that it contains so that they're available at the times that CAB code is expecting to see them is the reason for the startup tap-dancing I'm about to show you. All this code is generated for you by my SCSF.

   The CAB program contains its own CAB-based application class. The two uses of the "A" word have no relation to each other; this collision is an unfortunate overloading of nomen-clature. As we will see, the CAB application is created by the startup code. In turn, at the proper time, it creates the WPF application, which it stores in a member variable.

   The CAB application begins with a file called *Entry.cs*, shown next. It contains the Main method, which the Windows loader calls to start the program. It creates the CAB applica-tion and calls its Run method, thus:

```
public static class Entry
{
    [STAThread]
    public static void Main()
    {
        //start up the CAB application

        new ShellApplication().Run();
    }
}
```

   The CAB application class is shown on the facing page. Our new class *ShellApplication* derives from Boogaart's CAB base class *ApplicationShellApplication*, which is a CAB shell application that wraps (doesn't derive from, it contains) a WPF application. In its generic parameters, we pass it the class to use for the root work item (here, as is customary, a plain old *WorkItem*) and the class to use for the WPF application. This is represented by the class *App*, generated by SCSF in the shell project. *App*, in turn, derives from WPF *Application*, so you can customize your WPF application class. We usually don't do this, just as we don't usually customize our *WorkItem* class. We could pass plain old *Application*, as we pass plain old *WorkItem*, and it would work fine. I generate the separate *App* class files because that's the default behavior of the WPF project template in Visual Studio.

   The base class of the *ShellApplication* contains a Run method that performs the CAB pro-cessing that you are used to seeing, that is, starting up the object builder, enumerating and loading the modules. There is one important difference here in WPF. The CAB Run method internally calls the WPF application's Run method, which doesn't return until the app ter-minates. (This is the same as for Windows Forms.) However, the WPF application doesn't create its main window (the shell window) until this call, whereas in Windows Forms the main window had to be created before the call. Most existing CAB code expects the shell window to be created before the modules are enumerated and loaded so that modules can create views and place them into workspaces at module load time. Therefore, we override the AfterShellCreated method (in this case, Shell meaning Shell Application rather than Shell Form, as it previously did) and explicitly create the main WPF window there, instead of waiting for the usual WPF startup time.

The Shell *Application* class also contains a number of other method overrides (not shown). Notice that a number of SCSF services are added and removed, and the action catalog builder strategy is added.

```
public sealed class ShellApplication :
    ApplicationShellApplication<WorkItem, App>
{
    protected override void AfterShellCreated()
    {
        base.AfterShellCreated();

        // Create the main window. Do it through the RootWorkItem so
        // that it receives CAB processing, such as dependency
        // injection and placement of the Workspaces that it contains
        // into the RootWorkItem's Workspaces collection.

        ShellWindow sw = RootWorkItem.Items.AddNew<ShellWindow>();

            // Place the new main window into the WPF application, which
            // the base class has already created.

        Shell.MainWindow = sw;

        // Register user interface extension sites

        RootWorkItem.UIExtensionSites.RegisterSite(
                UIExtensionSiteNames.MainMenu, sw.MainMenu);
        RootWorkItem.UIExtensionSites.RegisterSite(
                UIExtensionSiteNames.MainStatus, sw.MainStatus);
        RootWorkItem.UIExtensionSites.RegisterSite(
                UIExtensionSiteNames.MainToolbar, sw.MainToolBar);

        // Show the main window

        sw.Show();

    }
```

2. The design of a SmartPart doesn't change very much from Windows Forms to WPF. The base class changes to the WPF ***System.Windows.UserControl***, which you can see on the facing page.

A WPF control always has an XAML file, containing XML data describing the structure of the control. This mechanism is conceptually quite similar to that used in Web Forms controls. The XAML for this control is shown here. Events that you want the control to fire to its container, such as the ***Loaded*** event, must be explicitly called out in the XAML as shown here. My SCSF generates this for you automatically.

```
<UserControl x:Class="RTHospitalWPF.PharmacyModule.PharmacyView"
    xmlns="http://schemas.microsoft.com/winfx/2006/xaml/presentation"
    xmlns:x="http://schemas.microsoft.com/winfx/2006/xaml"
    Loaded="OnLoaded" Height="Auto" Width="Auto"
  Background="{x:Static SystemColors.ControlBrush}">

  <!--other controls go here -->

</UserControl>
```

```csharp
[SmartPart]
public partial class PharmacyView :
 System.Windows.Controls.UserControl,
 IPharmacyView, ISmartPartInfoProvider
{
    public PharmacyView()
    {
        InitializeComponent();
    }

    /// <summary>
    /// Sets the presenter. The dependency injection system will
    /// automatically create a new presenter for you.
    /// </summary>

    private PharmacyViewPresenter _presenter;

    [CreateNew]
    public PharmacyViewPresenter Presenter
    {
        set
        {
            _presenter = value;
            _presenter.View = this;
        }
    }

    void OnLoaded(object sender, RoutedEventArgs e)
    {
        _presenter.OnViewReady();
    }

    #region ISmartPartInfoProvider Members

    public ISmartPartInfo GetSmartPartInfo(Type smartPartInfoType)
    {
        return new SmartPartInfo("Pharmacy", "PharmacyDescription");
    }

    #endregion
}
```

3. The code for creating a Smart Part and adding it to a workspace is identical to that used in Windows Forms. The *IWorkspace* interface methods, such as Show, have always been defined as requiring parameters of class *System.Object*, so they swallow the new WPF *UserControl* objects just as they used to swallow the Windows Forms *UserControl* objects. The code looks like this:

```
private void AddViews()
{
  // Create the view that we want to add

  IPharmacyView ipv =
        this.WorkItem.SmartParts.AddNew<PharmacyView>();

  // Add it to the workspace whose name we specify

  this.WorkItem.Workspaces[WorkspaceNames.RightWorkspace].Show (ipv);
}
```

4. The internal implementations of the workspaces have undergone the largest changes, although even here they are less extensive than the flashy colors in WinDiff would lead you to believe. Here is an example from the OnShow method of the *TabWorkspace* class. The old Windows Forms version is shown in the top excerpt and the new WPF version in the bottom excerpt. The main change is that the old *TabWorkspace*, based on the old *TabControl*, uses the class *TabPage* for its internal items, whereas the new WPF version use the new WPF class *TabItem*. You can also see that the old version hooks up a handler to be notified when the control fires its *Disposed* event, whereas the new one does not. The reason is that WPF controls don't provide or need the *IDisposable* interface. Windows Forms objects need it for releasing the scarce Windows GDI objects, such as HDCs and HWNDs, that they wrap. WPF objects don't contain these sorts of scarce resources and therefore don't need to release them.

```
<Windows Forms version>
```

```
protected virtual void OnShow(Control smartPart,
    TabSmartPartInfo smartPartInfo)
{
    PopulatePages();
    ResetSelectedIndexIfNoTabs();

    TabPage page = GetOrCreateTabPage(smartPart);
    SetTabProperties(page, smartPartInfo);

    if (smartPartInfo.ActivateTab)
    {
        Activate(smartPart);
    }

    smartPart.Disposed += ControlDisposed;
}
```

```
<WPF version>
```

```
public void OnShow(UIElement smartPart, TabSmartPartInfo smartPartInfo)
{
    PopulateTabItems();
    ResetSelectedIndexIfNoTabItems();

    TabItem tabItem = GetOrCreateTabItem(smartPart);
    SetTabItemProperties(tabItem, smartPartInfo);

    if (smartPartInfo.ActivateTab)
    {
        Activate(smartPart);
    }
}
```

5. The mechanism for negotiating shared user interfaces, which we saw in Chapter 5, "Shared User Interface Extension," hasn't changed much either. As before, the *UIElementAdapter* class provides a standardized bridge between an application's code and the shared user interface object that it wants to modify. The author of the WPF port has removed the *UIElementAdapter* class based on Windows Forms and written a new one that works with any object of class *System.Windows.Controls.ItemsControl*. The WPF classes *Menu*, *ToolBar*, and *StatusBar* all derive from this class.

You register *UIExtensionSites* as you always have, as shown in the following code sample:

```
protected override void AfterShellCreated()
{
    base.AfterShellCreated();

    //create the main window

    ShellWindow sw = RootWorkItem.Items.AddNew<ShellWindow>();
    Shell.MainWindow = sw;

    // Register user interface extension sites

    RootWorkItem.UIExtensionSites.RegisterSite(
        UIExtensionSiteNames.MainMenu, sw.MainMenu);

    RootWorkItem.UIExtensionSites.RegisterSite(
        UIExtensionSiteNames.MainStatus, sw.MainStatus);

    RootWorkItem.UIExtensionSites.RegisterSite(
        UIExtensionSiteNames.MainToolbar, sw.MainToolBar);

    //show the main window

    sw.Show();
}
```

6. The following code shows an example of adding items to a menu. It looks exactly like the Windows Forms–based menu extension shown in Chapter 5, except that you are creating WPF *MenuItem* objects instead of Windows Forms *MenuItem* objects. Fetching the *UIExtensionSite*, adding the objects to it, and hooking up the command invoker are identical. Thus:

```
private void AddMenuItems()
{
    // Fetch the UIExtensionSite that wraps the
    // main menu.

    UIExtensionSite uie = this.WorkItem.UIExtensionSites
        [UIExtensionSiteNames.MainMenu];

    // Create new menu items to be placed on the shared menu

    MenuItem PharmacyHeadMI = new MenuItem();
    PharmacyHeadMI.Header = "Pharmacy";

    MenuItem AddPrescriptionMI = new MenuItem () ;
    AddPrescriptionMI.Header = "Add Prescription";
    PharmacyHeadMI.Items.Add(AddPrescriptionMI);

    // User the UIExtensionSite to add the shared menu

    uie.Add (PharmacyHeadMI);

    // Hook up a command to be invoked when the specified
    // MenuItem fires the Click event

    this.WorkItem.Commands[CommandNames.AddPrescription].AddInvoker(
        AddPrescriptionMI, "Click");
}
```

F.   More Complex Example: Browser-Hosted XBAP Application

1. One additional advantage of WPF is its ability to be hosted in the Internet Explorer browser, a technique known as an XAML Browser Application (XBAP). When the user clicks on a link pointing to an XBAP application, the code is downloaded locally using Click Once deployment and then runs in the user's browser. One of the most vivid examples of this capability is the British Library's publication of medieval manuscripts, online at *http://ttpdownload.bl.uk/browserapp.xbap*.

Since WPF provides the capability of running in an XBAP, we also want our CAB and SCSF implementations to support that capability as well. It turns out not to be very hard to do. Martin Hoogendoorn (what is it with these double double-vowel guys?), who now works for Microsoft, published the initial release of this capability. I've taken his ideas and rolled them into an SCSF port to automate the process.

When you select an XBAP project from the guidance package, the generated project has a few small differences. First, the app must be (and is) generated with full trust because the CAB DLLs are not marked with the attribute that allows partially trusted callers. Second, the application class has to change a little bit. That code is shown on the facing page.

The main display area, the part that you've previously thought of as the shell form, is now a browser page, which derives from the class *System.Windows.Controls.Page*. The generated solution now contains an object of this class, represented by the class **Page1** in the sample code. At startup time, the CAB app doesn't show the main window in the previous way. Instead, it creates the main page. Then it hooks the application's *Startup* event notification, at which time it navigates to the main page, as shown on the facing page.

We also have to hook the application's *Exit* event to perform some cleanup, as shown at the bottom of the facing page. The CAB libraries formerly performed this cleanup for us, but in an XBAP application, they're not able to. Therefore the SCSF generates code to perform them at the proper time. I explain this condition on the next page.

```csharp
public sealed class XbapCabApplication : XbapApplication<WorkItem, App>
{
    // Main display area, used to be a window, now a Page

    private Page1 _mainPage;

    protected override void AfterShellCreated ()
    {
        base.AfterShellCreated ();

            // Create the main page. Do it through the RootWorkItem so
            // that it receives CAB processing

        _mainPage = RootWorkItem.Items.AddNew<Page1>();

            // Add handler that will navigate to main page when
            // application starts up

        Shell.Startup += new StartupEventHandler(Shell_Startup);

            // Add handler that will dispose of Root WI and Visualizer
            // when application shuts down

        Shell.Exit += new ExitEventHandler(Shell_Exit);

        // Register UIExtensionSites, and other startup code

        < code omitted … >
    }

    void Shell_Startup(object sender, StartupEventArgs e)
    {
        ((NavigationWindow)Shell.MainWindow).Navigate(_mainPage);
    }

    void Shell_Exit(object sender, ExitEventArgs e)
    {
        this.RootWorkItem.Dispose();
        if (this.Visualizer != null)
        {
            this.Visualizer.Dispose();
        }
    }
}
```

2. I had to make one small change to the CAB base libraries to support both WPF and XBAP cases from the same CAB library DLLs. I don't like changing these libraries for compatibility reasons, but this change is about as small as they get and I haven't found any good way around it.

The code for the method CabApplication.Run() is shown on the facing page. The call CabApplication.Start() maps down in the lower layers to the method Application.Run( ). In a Windows Forms or WPF application, this doesn't return until the main window is closed because it causes the CLR to start running a message pump from the user interface. In an XBAP application, the browser is already running its own message pump, so this method returns immediately.

The CAB library code was written for the former case and therefore disposes of the root work item and the visualizer after Start() returns, expecting this to be the end of the application session. An XBAP application will have just started running at this point, so we certainly don't want to wipe out the root work item. Therefore I've made the modifications shown in bold on the facing page. I've added a Boolean flag, which defaults to *true*. After Start**()** returns, I check this flag. If it's still *true*, then the disposal happens as before. (I've also added the protected variable *Visualizer* so that derived classes can dispose of this object.) Non-XBAP code doesn't need to change at all; it is disposed as usual. But an XBAP class sets this flag to *false* in its application code (shown here, generated by the SCSF). In this case, the CAB libraries won't do the cleanup, so the *Shell_Exit* event handler shown on the previous page has to. Thus:

```
public class XbapApplication<TWorkItem, TShell> :
   WindowsApplication<TWorkItem, TShell>
        where TWorkItem : WorkItem, new()
        where TShell : Application, new()
{
    // In constructor, set base class's flag to not dispose after Start
    // returns.

    public XbapApplication ()
    {
        this.DisposeAfterStartReturns = false;
    }

  < … rest of class >
}
```

```
/// Specifies whether or not the app disposes of the root work item
/// after Start returns. True in non-XBAP case, false otherwise

protected bool DisposeAfterStartReturns = true;

protected IVisualizer Visualizer;

public void Run()
{
  RegisterUnhandledExceptionHandler();
  Builder builder = CreateBuilder();
  AddBuilderStrategies(builder);
  CreateRootWorkItem(builder);

  Visualizer = CreateVisualizer();
    if (Visualizer != null)
        Visualizer.Initialize(rootWorkItem, builder);

  AddRequiredServices();
  AddConfiguredServices();
  AddServices();
  AuthenticateUser();
  ProcessShellAssembly();
  rootWorkItem.BuildUp();
  LoadModules();
  rootWorkItem.FinishInitialization();

  rootWorkItem.Run();
  Start();

    if (DisposeAfterStartReturns == true)
    {
        rootWorkItem.Dispose();
        if (Visualizer != null)
            Visualizer.Dispose();
    }
}
```

Chapter 8 Lab Exercises
CAB and WPF

1. Using the May SCSF, generate a project that is able to host WPF SmartParts. Add a WPF SmartPart to the project, and display it in a workspace.

2. Using my WPF SCSF and Boogaart's libraries, generate a new pure WPF application. Add a WPF SmartPart to the project and display it in a window.

Appendix
Generics

A.   Problem Background

1. The largest omission in the 1.1 Framework's object services is the inability to parameterize object types, as the template mechanism of C++ or Java allows. That might not mean much at first glance, especially to nonprogrammers, but here's what I mean and why it's important. Suppose I have an array, an object of class *System.Array*. That array is defined as holding objects of the universal base class *System.Object*, which means that it can hold any .NET object in existence. This capability is handy because we don't have to develop a special array to hold each class of object that we have to deal with. But it's rare that we use any particular array to hold objects of more than one class. Even though any element can hold an object of any class, we usually want all elements of an array to hold objects of the same class—strings, fish, birds, whatever. This means that storing an object of a different class, say a fish into an array of birds, probably represents an error in our program logic, even though the array class allows it. We'd like a way for the compiler to catch such errors. The facing page shows a code example.

In addition, this treatment of all types as *System.Object* requires a reference to each contained object. This, in turn, causes boxing of value types (automatically allocating space on the managed heap to convert them into reference types), a performance drag. Ideally, we'd like the compiler to know if the array was holding a value type (an object initially allocated on the stack and passed by value, such as ints, structs, and so on) so that our array could allocate the correct amount of storage to hold it by value instead of by reference. This would allow us to avoid boxing of value types, which consumes time and heap space and increases the frequency of garbage collections. Figure A-1 demonstrates the boxing and unboxing that takes place when we use a *System.Array* to hold value types.

The only way we can currently accomplish these goals is by writing our collection classes for each type we might want to hold, or at least placing wrappers on every method on the existing array class. This effort would entail a large amount of repetitive infrastructural development work. And if there's one thing you should have learned from everything I've ever written about .NET, it's that large amounts of repetitive infrastructural work belong in the operating system, not your program logic. We would like a mechanism that would allow us, when we create an array, to pass it the type of object that we want it to hold and have it reject attempts to make it hold anything else. We would like to somehow tell it, "Hey, I know you can hold anything, but in this case, I want you to restrict yourself to holding only birds." Since the compiler would know the class that the array holds, we like to assign an object we fetch from it directly into a variable of the correct type without needing to cast it.

Developers of this super-smart array class would like to write the code only once and have it magically work with any class that the client tells it to. And we'd like to easily use this mechanism in writing any class, not just an array, that needs to hold another potentially varying class.

```
private void button1_Click(object sender, EventArgs e)
{
    // Create a standard ArrayList

    ArrayList al = new ArrayList();

    // Create an integer, store it in the ArrayList
    // Boxing takes place

    int i = 5;
    al.Add(i);

    // Create a color, store it in the ArrayList
    // The weak typing of the ArrayList allows this,
    // but it is most likely a logic error

    Color c = Color.AliceBlue;
    al.Add(c);

    // Fetch the int from the ArrayList.
    // Cast is required, one more source of possible error
    // Unboxing takes place

    int j = (int)al[0];

}
```

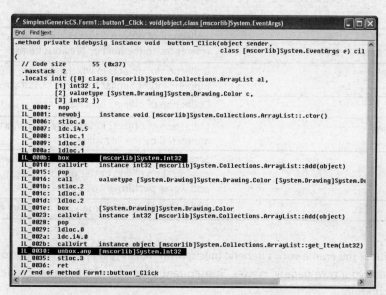

FIGURE A-1 Intermediate language disassembly showing boxing operation when value types are added to non-generic ArrayList.

B.   Solution Architecture

1. Version 2.0 of the .NET Framework supports parameterizing classes by a mechanism called *generics*. This mechanism looks and feels very much like the template mechanism in C++ or Java. The designer of a class, such as *System.Collections.Generic.List*, specifies that the class can accept a type as a parameter by using a special syntax, as shown on the facing page. Note that IntelliSense has properly detected the type that the generic list holds (in this case, string) and displays it, thus being more specific and therefore more useful.

   The client programmer is passing the String type, saying, "Here, create me a list that holds only strings, please." Any attempt to store an object other than a string in that list causes a compiler error, and it eliminates the casting when you fetch an entry from it. It also eliminates the boxing and unboxing of value types. It works properly across all languages and makes IntelliSense prompt more useful. The first example of this appendix demonstrates this generic mechanism.

   The Framework provides a set of collection classes that use this mechanism. You can find them in the namespace *System.Collections.Generic*. Some of the more useful classes in this namespace are described in Table A-1.

TABLE A-1 Generic Classes provided in .NET Framework 2.0

Class	Description
Dictionary	Collection of objects indexed by keys.
LinkedList	Collection of objects in which items are accessed by references to the next or previous items.
List	Collection of objects in which items are accessed by numeric index. Similar to *ArrayList*, but strongly typed.
Queue	Collection of objects organized as first-in, first-out.
SortedDictionary	Collection of objects indexed by keys, in which objects are sorted by the key value.
SortedList	Similar to *SortedDictionary*, but with different internal organization giving better performance in some cases and worse performance in others.
Stack	Collection of objects organized as last-in, first-out.

In addition, the Framework's generic mechanism makes it very easy for you to design and work with your own generic classes. Subsequent examples of this appendix demonstrate this technique.

C# syntax showing creation of new generic *List* holding strings:

```
private void button3_Click(object sender, EventArgs e)
{
    // Create a new generic List, telling it to accept only objects
    // of class String

    System.Collections.Generic.List<String> MyStringList =
        new System.Collections.Generic.List<String>();

    MyStringList.Add(
```
```
void List<string>.Add (string item)
item:
    The object to be added to the end of the System.Collections.Generic.List<T>. The value can be null for reference types.
```

FIGURE A-2 IntelliSense displaying help for generic class.

VB Syntax showing creation of new generic *List* holding integers:

```
'Create a new generic List, telling it to accept only
' objects of class int

Dim l As New System.Collections.Generic.List(Of Integer)
```

C.   Simplest Example

1. The facing page shows an example of using a generic collection to obtain the benefits of strong typing and avoid the performance drag of boxing. I first instantiate an object of the class *System.Collections.Generic.List*, which is an expandable array, the generic equivalent of *ArrayList*. In its angle brackets I pass the type that I want it to hold, in this case, *int*. The developers of this class have written it in such a manner as to check the type that it is passed and allocate storage for holding its value, so boxing and unboxing don't take place, as you can see in Figure A-3. (We see these techniques in the next example.)

   The compiler now knows that this list accepts only integers, so storing an integer in it is allowed, as in the second paragraph of code, but storing another type in it, such as a color, fails. That's why the line attempting to store the color in the *List* is commented out; it wouldn't compile otherwise.

   When we fetch the integer out of the *List*, we no longer have to cast it from *Object* into the *int* type. The compiler, again knowing the type that this list contains, sees that we are assigning it to the correct type and allows the code to compile.

   It's a win-win-win situation all around, except for the guys who have to write the generic *List* class, which (as we'll see) is slightly trickier than writing a nongeneric class. But even here, you gain large productivity benefits by now having to write a strongly typed collection class for every type you might want to hold. (Anyone remember *CIntArray* and *CCharArray* and their brethren from the MFC?)

   In VB, the type parameter is passed using the *Of* operator in parentheses, thus:

```
Dim 1 As New System.Collections.Generic.List(Of Integer)
```

```
private void button2_Click(object sender, EventArgs e)
{
    // Create a new generic List, telling it to accept only
    // objects of class int

    System.Collections.Generic.List<int> l =
            new System.Collections.Generic.List<int>();

    // Create a new int, add it to the list.
    // Boxing does NOT take place

    int i = 7;
    l.Add(i);

    // Create a new color

    Color c = Color.AntiqueWhite;

    // Attempting to add the color to the list that accepts only ints
    // causes a compiler error. This line does NOT work.

    // l.Add(c);

    // Fetch the int from the ArrayList
    // Cast is not needed
    // Unboxing does not take place

    int j = l[0];

}
```

FIGURE A-3 Intermediate language disassembly showing that boxing does NOT take place when value types are added to generic list.

D.   More Complex Examples: Writing Our Own Generic Classes

1. It is simple to write our own generic classes. In our class declaration, we specify a parameter type name in angle brackets, as shown on the facing page. It is customary to use very short names for these, such as T and K, but I've gone with a longer one to remind you constantly of the type that name represents. In VB, the type parameter is declared using the *Of* operator. You then use that type name parameter to represent whatever type the user passes at instantiation time.

chuck: come back

> **Note** Generics work only for parameters whose type is known at compile time. They do not automatically perform reflection to figure it out at runtime. They work very well, and are a big advantage, when dealing with types you know about. They aren't as useful for types you don't discover until runtime.

In VB, the class declaration is

```
Public Class GenericHolder(Of TypeMyClientToldMeToHold)
```

And the member variable declaration is

```
Private objectThatIHold As TypeMyClientToldMeToHold
```

```
public class GenericHolder <TypeTheClientToldMeToHold>
{

    // At construction time, show a dialog box
    // Use the parameterized type we are passed to
    // demonstrate to the user that we know what it is.

    public GenericHolder()
    {
        MessageBox.Show("Type I hold is: " +
                typeof(TypeTheClientToldMeToHold).ToString()) ;
    }

    // This member variable magically holds whatever type
    // the client has specified at instantiation time

    private TypeTheClientToldMeToHold objectThatIHold;

    // These methods magically accept and return the parameterized
    // type the client has specified at instantiation time

    public void Add(TypeTheClientToldMeToHold obj)
    {
        objectThatIHold = obj;
    }

    public TypeTheClientToldMeToHold Fetch()
    {
        return objectThatIHold;
    }
}
```

2. You can have a lot of fun with generic classes. For example, you can design a class to accept more than one parameterized type. The *Stack* and *Queue* classes accept only one,
thus:

```
public class Stack <T>
```

But the *Dictionary* is declared as accepting two different ones, the values that it holds and
the key by which the values are indexed. Note that they are listed in an order I find coun-
terintuitive. Thus:

```
public class Dictionary <TKey, TValue>
```

This declaration allows you to write methods such as *Dictionary.Add*, thus:

```
public void Add(TKey key, TValue value)
{
    this.Insert(key, value, true);
}
```

3. Snag: You can't use the == operator on parameterized types because they could be anything and the compiler doesn't know whether they support that operator. So the *Dictionary* class can't just do this:

```
public TValue GetValue (TKey DesiredKey)
{
    for (int i = 0 ; i < this.KeyArray; i++)
    {
        if (this.KeyArray[i] == DesiredKey)  // this line won't compile
        {
            return this.ValueArray [i] ;
        }
    }
}
```

One technique you could use is to place a *derivation constraint* on the generic class. This is a way of specifying that the parameterized type must derive from a specific base class or implement one or more specified interfaces. You could say, for example, that the key class has to implement the interface *IComparable* so that you can use it for comparisons. You specify such a derivation constraint thus:

```
public class Dictionary <TKey, TValue> where TKey : IComparable
```

Better yet, you can use the generic version of the *IComparable* interface, get type safety, and avoid boxing, thus:

```
public class Dictionary <TKey, TValue> where TKey : IComparable<TKey>
```

Your comparison class would then do this:

```
public TValue GetValue (TKey DesiredKey)
{
    for (int i = 0 ; i < this.KeyArray; i++)
    {
        if (this.KeyArray[i].CompareTo(DesiredKey) == 0)
        {
            return this.ValueArray [i] ;
        }
    }
}
```

If a programmer attempts to pass a parameterized class that does not comply with the derivation constraint, he gets a compiler error.

Note that the *Dictionary* class does not use this technique; instead, Microsoft has done all kinds of stuff internally to keep you from having to think about that.

If you do use a derivation constraint, the compiler allows you to cast the parameterized type into this type, but otherwise you cannot cast generic types other than to *System. Object* because the compiler does not know whether they are supported.

When the generic class contains more than one parameterized type, you can place derivation constraints on any or all of them.

4. If your generic class creates a new instance of the parameterized type, you have to specify that this type contains a default constructor, that is, one with no parameters. Otherwise, the compiler cannot know at compile time whether the construction would succeed. For example, suppose you wanted to do this:

```
public class SomeClass <TValue>
{
   public TValue CreateNew ( )
   {
        return new TValue( ) ;            // this won't compile
   }
}
```

This code does not compile because the compiler can't guarantee that it will be able to call the default constructor on the parameterized type. However, you can constrain the parameterized type so as to require it to contain a default constructor, thus:

```
public class SomeClass <TValue> where TValue : new ( )
{
   public TValue CreateNew ( )
   {
        return new TValue( ) ;
   }
}
```

And, of course, you can combine this constructor constraint with a derivation constraint, thus:

```
public class SomeClass <TValue> where TValue : new ( ), IComparable
{
   public TValue CreateNew ( )
   {
        return new TValue( ) ;
   }
}
```

5. You can, if you want to, constrain a parameterized type to be a reference type or constrain it to be a value type. In the former case, you would write

```
public class SomeClass <TValue> where TValue : class
```

and in the latter case, you would write

```
public class SomeClass <TValue> where TValue : struct
```

The latter option is used in the class *System.Nullable <T>* class, to which we now turn our attention.

E.   An Interesting Use of Generics in the .NET Framework

1. One of the more interesting applications of generics in the .NET Framework is the **Nullable<T>** class. It is easy to know when a variable of a reference type does or doesn't refer to a valid object. In the latter case, its value is null. But this isn't the case with value types because zero is an entirely valid and reasonable value.

   Consider, for example, the case of parsing an XML document, where you have an element defined in the document schema as containing an double. Suppose further that the document schema defines that element as optional; in other words, it may appear either once or not at all. In a .NET wrapper class that complies with the schema, how would you know whether that optional element had or had not appeared? Today, you have to handle that with a separate Boolean variable—for example, *double NetWorth* and *bool NetWorthSpecified*. It's annoying, it means that you have to be constantly thinking what a reference type is and what a value type is, and it's just a constant burr under your blanket. And when you have a few hundred of these things in a schema, the hassle adds up quickly.

   The generic class *System.Nullable* represents a wrapper for any value type. It contains an internal variable called *HasValue*, which tracks whether the internal value type has been set. It starts out false. You use it just like a value type, except that it recognizes null (*Nothing* in VB) as a special case value different from the number zero, sort of like a tristate ransistor (ITL) if you remember those (values *0*, *1*, or *off*). You can set it to and compare it to null. A code sample is shown on the facing page.

```
// Declare a variable that is a nullable integer. It starts out
// with the value null

private Nullable<int> ni ;

private void button4_Click(object sender, EventArgs e)
{
    // Compare it to null. If so, inform user

    if (ni == null)
    {
        MessageBox.Show("ni is null");
    }

    // If not, treat it like an integer

    else
    {
        MessageBox.Show("ni value = " + ni.ToString());
    }
}

// Set its value. If it is null at this point, it
// will no longer be

private void button5_Click(object sender, EventArgs e)
{
    ni = Convert.ToInt32(textBox1.Text);
}

// Set its value back to null

private void button6_Click(object sender, EventArgs e)
{
    ni = null;
}
```

Appendix A Lab Exercises
Generics

1. Open the solution in the folder \Generics\Lab\Template, from the Appendix A sample code from the book's Web site. Build it and run it just to make sure that nothing silly is broken.

2. The program performs a performance test with both a value type and a reference type. The template already does this with a nongeneric stack. Add code that performs the same test with a generic stack. Observe the performance difference with both value and reference types.

3. The results of that test may not be exactly as you expect. Dig into the *System.Dll* using ILDASM and try to figure out why.

4. Run the performance test again with larger value types and see what sorts of numbers you get. Try creating your own extra-large value type, perhaps a struct containing 25 doubles, and see if that affects the performance any. Try the reference type performance test with long strings versus short ones and see how that affects the test.

5. Try the performance tests again where you start timing the operation BEFORE you create the stack rather than after. How does this affect performance?

6. Try writing your own generic class *MyOwnStack*, containing methods Push and Pop. For the purposes of this class, you can use a simple array for internal storage. Try the performance tests with this stack.

7. Extra Credit: Try creating a rudimentary generic class called *MyOwnDictionary*, accepting parameterized types as both keys and values, containing methods Add and Get. For the purposes of this exercise, you can use arrays for internal storage. Experiment with different ways of performing the comparison to the key value in the Get method and think about the advantages and disadvantages of each.

Index

Symbols

= = operator, 189

A

action, defined, 140
[Action] attribute, 143
action catalog service, 40, 42, 153
  background, 139
  builder strategy, 167
  diagram, 141
  My Own Time Service, 142–147
  passing and modifying parameters, 148–149
  solution architecture, 140–141
action condition object, 140
Action method, 68
ActionCatalogService, 18, 40, 42, 66
Activate method, 80, 82
ActivateTab property, 94
Active X controls, 22
ActiveSmartPart, 80, 158
Add event, 118
Add method, 110, 114
Add < >, 59
<add> element, 49
AddBuilderStrategies method, 146
AddInvoker method, 106, 108
AddNew method, 160
AddNew< > method, 59, 62, 64
AddOnDemand attribute, 51
AddPublication( ) method, 132
AddServices method, 49, 147
AddServices( ) method, 44
AddSubscription( ) method, 134
AddUserInterfaceExtensions, 118
AfterShellCreated method, 37, 167
AfterShellCreated( ) method, 102
App class, 166
app.config file, 40, 44, 49
AppDomain, 56
application catalog service, 40
Application.Run( ) method, 176
applications
  browser-based vs. client, 2
  client. *See* client applications
  composite, 98
  monolithic, 7, 8
  shell. *See* shell applications
  Walkthrough Quickstart, 20–29
  Windows Forms, 11
ApplicationShellApplication, 166

ApplySmartPartInfo method, 80
architectural teams, 6
architecture
  document view, 28
  modular, 18
  Module—View—Presenter (MVP), 24, 26
  pluggable, 6
  solution. *See* solution architecture
  three-tier, 84
ArrayList, 184
arrays, 180, 184
asynchronous occurrences, 123
Authenticate method, 44
authentication, 6
authentication services, 44, 46–49
authenticity identity token, 46

B

Background option, 126
Boogaart, Kent, 162
Boolean flag, 176
Boolean parameters, 140, 148
Boolean variables, 192
British Library, 174
browser-hosted XBAP application, 174–177
bugs
  bugger tracing, 30–33
  programming, 6
Buildup method, 44
business logic
  CAB, 17, 62, 84, 139
  Excel, 13
Business Module, 52

C

C++, 180, 182
C# syntax, List holding strings, 183
CAB. *See* Composite UI Application Block (CAB)
CabApplication class, 61
CabApplication.AddRequiredServices, 47
CabApplication.LoadModules, 64
CabApplication.Run( ), 176
CabApplication.Start( ), 176
CanExecute method, 140, 146
catalog provisioning, 6
catalog service, 40, 42
CCharArray, 184

child vs. parent chain, 60
CIntArray, 184
Class Wizard, 17, 163
classes. *See also specific classes*
  classic CAB, 20
  generic, writing, 186–191
  parameterization, 182
  SCSF, 18
  write-only, 22
ClickOnce, 16, 174
client applications
  architectural team requirements, 6
  background, 2–7
  dynamic link libraries (DLLs), 14–16
  loose coupling with CAB, 8–16
  operation team requirements, 6
  programmer requirements, 6
  Smart Client Software Factory (SCSF), 17–19
  tracing and visualization, 30–33
  user requirements, 6
  walkthrough, classic CAB, 20–29
Close method, 80
Close( ) method, 88
code
  CAB library, 14, 176
  EntLib removal, 39
  EventTopic access, 128
  good vs. bad, 18
  handler, 26
  menu additions, 106–107
  object building, 142
  presenter, 26, 89
  services, 47, 50
  SmartPart creation, 170
  view, 28, 81, 87
  Visualizer object, 162
  workspace, 55
  WPF support, 156
  writing, 18
  XBAP, 16
collections, WorkItem class, 58–61
COM framework, 98
COM+ event system, 123
command handlers, 68
Command objects, 58, 108
CommandAdapter class, 162
CommandHandler attribute, 108
Commands collection, 106, 108
CommandStatus, 108
common language runtime (CLR)
  message pump, 176
  versioning issues, 6

David S. Platt

David S. Platt teaches .NET and CAB programming at Harvard University Extension School and at companies all over the world. He is the author of ten programming books prior to this one. His *Introducing Microsoft .NET* from Microsoft Press introduced thousands of programmers to that environment. Even today, 4 years after its most recent release, it is outselling Tom Clancy's *Every Man a Tiger* on Amazon.com, which tells you what kind of geeks buy their books there. His magnum opus, *Why Software Sucks* (Addison-Wesley, 2006, *www.whysoftwaresucks.com*), points out ways in which software MUST improve if it's to accompany humanity into the twenty-first century.

He is famous for his engaging presentation style. "He's the only guy I know that can actually make a talk on COM's apartment threading model funny," said one student. Microsoft named him a Software Legend in 2002.

Dave holds the Master of Engineering degree from Dartmouth College. He did his undergraduate work at Colgate University. When he finishes working, he spends his free time working some more. He wonders whether he should tape down two of his daughter's fingers so she learns how to count in octal. He lives in Ipswich, MA.

Nickname: "The Mad Professor"

Favorite Web site: radiomargaritaville.com

Comment most frequently elicited from children at adjoining breakfast restaurant tables before he's had his morning coffee: "Mommy, what's wrong with that man?"